Genesis: The Explanation

What Really Happened "In the Beginning"

T. Adams Upchurch, PhD

An Imprint of
GlobalEdAdvancePress

GENESIS: THE EXPLANATION

What Really Happened "In the Beginning"

Copyright © 2016 by T. Adams Upchurch

Library of Congress Control Number: 2016941116

Upchurch, Thomas Adams 1964 —

ISBN 978-1-935434-83-2

Subject Codes and Description: 1: REL106000: Religion: Religion & Science; 2: REL006630 Religion: Biblical Studies - History & Culture 3: SCI015000 Science: Cosmology

All rights reserved, including the right to reproduce this book or any part thereof in any form, except for inclusion of brief quotations in a review, without the written permission of the author and GlobalEdAdvancePRESS.

Except where otherwise noted: Scripture taken from the New King James Version, Copyright 1982 by Thomas Nelson, Inc. Used by permission. All rights reserved.

Cover design by Brian Lane Green

Printed in Australia, Brazil, France, Germany, Italy, Poland, Spain, UK, and USA Also available on Espresso Book Machine and anywhere good books are sold.

The Press does not have ownership of the contents of a book; this is the author's work and the author owns the copyright. All theory, concepts, constructs, and perspectives are those of the author and not necessarily the Press. They are presented for open and free discussion of the issues involved. All comments and feedback should be directed to the Email: [*comments4author@aol.com*] and the comments will be forwarded to the author for response.

Order books from www.gea-books.com/bookstore/ or any place good books are sold.

Published by
Post-Gutenberg Books
An Imprint of
GlobalEdAdvancePRESS

In memory of
Susan Cox
1961-2015

See ya in Heaven someday, lil' big sis.

Genesis: The Explanation

What Really Happened
"In the Beginning"

Contents

Publisher's Preface	9
Foreword: E. Basil Jackson, MD, PhD, JD	11
Conceptual Introduction	17
THE PUZZLE ANALOGY	
THE BIBLE	
Prequel: Before The Beginning	31
THE INDESCRIBABLE ONE	
THE LOVE STORY	
THE CELESTIAL DRAMA	

1. **GENESIS Chapter I** — 65

   A DAY IS AS A THOUSAND YEARS

       MILLENNIUM ONE
       MILLENNIUM TWO
       MILLENNIUM THREE
       MILLENNIUM FOUR
       MILLENNIUM FIVE
       MILLENNIUM SIX

2. **GENESIS Chapter II** — 103

   THE LABORATORY OF EDEN

       MAN OF DIRT
       ADAM'S RIB

3. **GENESIS Chapter III** — 129

   TO DIE FOR

       THE CHOICE
       THE CONSEQUENCES
       THE CURSE

4. **GENESIS Chapter IV** — 165

   SURVIVING THE STONE AGE

       FAMILY LIFE
       THE FIRST DEATH
       EAST OF EDEN
       THE FIRST POLYGAMIST
       THE FIRST PRAYERS

GENESIS: THE EXPLANATION

5. GENESIS Chapters V - VII   211
    AS IT WAS IN THE DAYS OF NOAH
- THE FIRST ASTRONAUT
- THE FIRST SAVIOR
- THE GREAT CORRUPTION
- THE GREAT DIGRESSION
- THE GREAT PREPARATION

6. GENESIS Chapters VIII - XI   253
    OF RAINBOWS AND REGENERATION
- EARTH DIVIDED, HUMANITY UNITED
- HUMANITY DIVIDED, EARTH OVERSPREAD

CONCLUSION   277
- RECAPPING HOW IT ALL REALLY HAPPENED

BIBLIOGRAPHIC ESSAY   283
- **METHODS OF RESEARCH AND CITATION**
- **SELECT SOURCES CONSULTED**
  - GENESIS AS HISTORY VERSUS MYTH
  - SECULAR SCIENCE VERSUS RELIGION
  - THE ATHEIST CASE AGAINST GOD AND CREATIONISM
  - CREATIONISM VERSUS GODLESS EVOLUTION
  - JEWISH INTERPRETATIONS OF GENESIS
  - UNORTHODOX ONTOLOGICAL ARGUMENTS
  - WHO WROTE GENESIS?
  - SATAN
  - HEAVEN
  - CHRISTIAN INTERPRETATIONS OF GENESIS
  - GNOSTICISM
  - SCIENCE, GENERALLY
  - PREHISTORIC ARCHAEOLOGY AND ANTHROPOLOGY
  - THE ANCIENT ALIEN THEORY
  - MISCELLANEOUS, RANDOM, AND WORTHY OF MENTION
- **SOME ADDITIONAL SOURCES CONSULTED**

Author's Notes: Purposes, Caveats, And Disclaimers   371
- BRIDGING THE GAP FROM ACADEMIA TO MAIN STREET
- BEING WISE AS A SERPENT BUT HARMLESS AS A DOVE

Acknowledgements   382

About The Author   383

Publisher's Preface

THE PUZZLE BOX PICTURE

On the puzzle box of Genesis, there was no four-color picture to guide putting the pieces together. Dr. Upchurch, a tenured Professor of History in the University System of Georgia, has provided some of the missing pieces and a possible picture for the box. His "Explanation" of what really happened in the beginning is a worthy addition to the body of literature that interprets the Beginning as presented in Genesis.

Dr. Upchurch is a prolific writer of academic books based on his historical interest in what happened in Early America. Now he turns his talent to what happened "In the Beginning." Readers interested in what happened in the early chapters of Genesis will appreciate this explanation. With the perspective of a historian, rather than a theologian, Dr. Upchurch brings his critical thinking skills to a general audience to facilitate a discussion of what really happened in the beginning. This work began as material for a Critical Thinking class, then after putting the original thinking about the

Genesis issue on paper the author identified the four corners of the puzzle: (1) the Genesis creation story, (2) the secular theory of biological evolution, (3) the ancient pagan mythology, (4) and the modern theories about extraterrestrial intelligence. This conceptual framework gave this book a distinctive place in the body of literature on Genesis.

Starting with the four corners of the puzzle, Dr. Upchurch combined Biblical, scientific, historical and parapsychological knowledge into an explanation of the mystery of existence to create a word picture for the puzzle box. Hopefully, those who consider The Explanation will get a practical understanding of what really happened in the beginning. Dr. Upchurch looks to the Judeo-Christian Bible as a primary source to assist the construction of his picture on the puzzle box. It is difficult for one person to get each and every detail exactly the perfect color and shape that would please each critical eye, but Dr. Upchurch brings fresh colors and shapes to the discussion. All who read this work will find it both exciting and inspirational. Readers would be better informed of The Explanation by closely observing the author's Conceptual Introduction, Prequel: Before the Beginning and the Conclusion and Sources. Read, learn, and enjoy!

— *Hollis L. Green, ThD, PhD, DLitt*

Foreword

A fascinating book because it is imaginative and cognition stretching — all stimulated by an author who loves The Book.

It is usually worthwhile on opening a book to ask why the author chose to record his thoughts on paper. In this case the author has provided us with a luculent answer: *"The main goal of this book is to explain to open-minded unbelievers, skeptics and agnostics (with emphasis on open-minded) why the overarching message of the Bible is true, showing how it is or at least can be believable."*

If we can be sufficiently open-minded and with a modicum of psychological maturity to accept the author's statement as coming from an honest and ethical scholar, with a profundity of knowledge and sagacity, and a love for the Bible as the supernaturally inspired revelation from God, then we will be in a good position to read, learn and inwardly digest and perhaps disagree with some of the ideas of the author. Analysis of the author's explanatory statement will tell us something about the work and about the author.

- He desires to encourage non-believers to think;

- He believes that the Bible is true;
- He would like to demonstrate to open-minded readers that the Bible is at least credible. A lofty goal indeed!

Of course I must be transparent and admit that, having read a plethora of works on Genesis, I was excited to see any new work on this subject. My experience has been that, having read a library of books on Genesis during a long life, I had come to the conclusion that anything on Genesis published today by evangelical authors would likely turn out to be evangelical regurgitation. Here I was wrong!

Criticism of the author and his ideas should be expected and are sure to come! In my opinion, even though the author has directed his thoughts and ideas primarily to the unbeliever, skeptic and agnostic, it is more likely that the bulk of criticism which will constitute 'incoming' artillery directed at him will be from Bible accepting evangelicals. The author gives evidence that his motivation is pure, as is his love for God's word, but he must be aware that as he pursues his goal he will be crucified by some from both sides of the religion-science controversy. I am also sure that he already knows that wounds inflicted by friends are the most painful and yet in his determination to search for an Explanation he is willing to take the risk. He is to be admired!

It has been a long time since after completing the first reading of a book I returned immediately to the first page and started all over again to make sure that I had understood at least some of the content which I had found to be fascinating, erudite, fantastic and mind-stimulating. That is, in my opinion, the highest order of appreciation I could give to any erudite and thought provoking work. I believe it would be fair to anticipate that a significant portion of the expected criticisms of Professor Upchurch is that he has fallen prey to the seductive attractions of science fiction. He will also be criticized vehemently for even mentioning the word 'evolution' but at times his critics will be those who have not understood the difference between micro and macro evolution.

Ab initio, on my first introduction to this book my mind flashed back to Shakespeare's Henry Vth. and words that I had not recalled since my high school years.

> *Now entertain conjecture of a time*
> *When creeping murmur and the poring dark,*
> *Fills the wide vessel of the universe.*

Words that might have been used to describe the disruption of Genesis 1:1-2. Later as I continued to read I was surprised, and yet not surprised, to read the very word 'conjecture' used by the author.

The author, obviously reflected in his many years of scholarship and perennial contemplation, is willing to take the risk of considering 'other possibilities' and other 'Explanations'. The fortitude to take such risks requires both personal security and academic guts! I have on occasion thought of such possibilities but I doubt if I would have had the guts to air such contemplations publicly! However, I have the feeling that this type of free unhindered and unfettered thought may at times do more for theological alignment than a lifetime of ossified theological regurgitation.

He understands the significance of the psychological aspects of the term 'Fundamentalism' better than many evangelical authors. Let me explain. The word fundamentalism is commonly used incorrectly among Christians and literalist Bible scholars and students. To such it means a concrete rigid literal interpretation of the words of the inspired book and giving little notice to the importance of a variety of aspects of literature, such as genre, figures of speech etc. It means accepting without question a series of dogmatic assertions and opinions which often appear to be based on a Pontifex Maximus authoritarian pronouncement rather than anything associated with critical thinking or reasonable contemplation. This is often an a-theological theology because it reflects the masquerading of rigid concrete psychopathology in theological language! Reading this work opens up new vistas in the understanding of a written narrative.

Upchurch cannot be accused of making dogmatic assertions and he fully understands the difficulty of getting one's arms around these ideas, speculations, conjectures and concepts. The evidence that the author is an honest and secure and erudite thinker is the fact that he regularly claims that his ideas are mere tentative suggestions and must never be tainted with the defensive dogmatism.

The author has an amazing breadth of knowledge in a plethora of fields and disciplines. It is refreshing to note that the author, in spite of and perhaps because of his academic experience, remains secure in his Christian beliefs and the authority of the Inspired Word. His love for the book and for the Lord is clearly reflected on page after page. He is never dogmatic regarding his own tentative assumptions and ideas as he raises possibilities for consideration. From Upchurch I have been reminded that it is no heresy to think, reason and consider other possible explanations. At times exegesis may be illuminated by eisegesis!

The book reminds me of my training and experience as a classical psychoanalyst. Having studied Freud for many years I do not 'believe' everything he says, but I am very happy to use some of his ideas and concepts when they help me to explain clinical phenomenology. In the same way Professor Upchurch tentatively offers suggestions which may be somewhat useful in his ontological search but he is careful to emphasize that he does not 'believe' as dogma that which has not been revealed.

The author must anticipate criticism, from those with ossified cognition, for he is entering the "No enter areas" of 'cognition not permitted'. He will be criticized, by those who do not understand his purpose or who are psychologically threatened by his attempt to engage in an imaginative flight into the fantasy world of science fiction. They fail to understand his attempt to consider how it might have been! On first superficial reading the work begins to sound like science fiction until suddenly the reader realizes that some of his ideas just may be consistent with the Hebrew text. Such skillful intermingling of established exegesis and fantasy/conjecture amounts at times to sanctified imagination. Let the critics remember that science fiction has developed into an acceptable genre in its own right. The authors of such tend to engage in a proleptic view of the future or the metamorphosis of the immediate present. Professor Upchurch, however, uses elements of the genre to EXPLAIN things which occurred in the past and he is careful to notify his readers that this is his purpose. One should keep in mind that science fiction is not a legitimate theological tool but that exegesis is rarely bereft of a modicum of eisegesis.

His thoughts on the Nephilim and the angels who lost their first estate are fascinating and amazingly consistent with some of the archeologic and skeletal evidence recently studied in Peru and elsewhere. In the opinion of this reviewer a detailed study of *katabole* in the New Testament and the LXX will tend to support some of the author's conjecturers and conclusions. The author is clear about his conclusions: *"Christian theology seasoned with the right amount of secular science and out-of-the-box-thinking allows for the best, if not the only reasonable, explanation for the mysteries of life and the universe. In other words, to more of the hard questions than anything else does, at least in my opinion".*

A fascinating, exciting, mind-stretching book from which everyone, believer and unbeliever, will benefit!

— E. Basil Jackson, MD, PhD, JD
Distinguished Professor of Psychiatry, Medicine, and Law

Conceptual Introduction

THE PUZZLE ANALOGY

In a typical jigsaw puzzle, there is a box with all the pieces inside and a picture of the finished product on the outside. To put a puzzle like that together requires a lot of trial and error; there is no better way than to dump all the pieces on a table and begin picking through them looking for parts that match up. Scattered randomly among the hundreds of odd pieces are the four corner pieces and a few dozen side pieces. Finding and matching those pieces together first is generally the best strategy. Once they have been assembled into an outer frame, the inner pieces will eventually come together to complete the puzzle and form the picture on the box.

This book began as an effort to piece together something like a mental jigsaw puzzle. In this analogy, the box represents my mind, and the picture on the box is everything I see when I look around—the earth and sky and outer space, the plants and animals and the human race; in other words the physical universe and all that is within it. To solve this puzzle would be to figure out how it all started and why it is here. To show you the finished product after all the pieces are assembled would be to give you *Genesis: The Explanation* (which will be called simply *The Explanation* from here on). I started by dumping hundreds of pieces of information out of my mind and onto paper and then identifying the four corner pieces: the creation story in the book of Genesis, the secular theory of biological evolution, ancient pagan mythology, and modern theories about extraterrestrial intelligences. Starting with them, I then matched dozens of side pieces that form the four sides of the outer frame: Biblical knowledge, scientific knowledge, historical knowledge, and parapsychological knowledge. From there, by trial and error, the inner pieces came together in my mind to make the picture on the box. The finished product that you will eventually see here is thus *The Explanation* for the mystery of existence.

The process of piece-fitting has been arduous, and it is worth discussing here at the outset. Everyone who has ever tried their hand at solving this puzzle of the mystery of existence knows that there are only two realistic possibilities about how the picture we see around us could have been made. One is the naturalist/materialist way, and the other is the supernatural/metaphysical way. There is no middle ground between them. The former says the physical universe has simply always existed, or else it came into being by natural causes without the help of a God-character, whether Yahweh, Jesus, Allah, or whatever the various religions choose to call their creator deities. Proponents of this possibility believe that, given enough time for research, science will eventually be able to explain how we ended up with this particular picture on the box. Within this materialist camp are basically two diverse schools of thought, which might be dubbed the "earthbound" school and the "otherworldly" school, for lack of anything better to call them. The one says that life evolved on earth with no outside interference or influence. The other says life evolved somewhere else in the universe first and then was brought here either deliberately by extraterrestrial beings (who are themselves products of purely materialistic evolution eons ago) or accidentally by a comet, asteroid, or similar celestial body which just happened to be passing through our galactic neighborhood. This is the theory called "pan-spermia."

The supernatural/metaphysical way says that something exists outside of the physical universe which is beyond our ability to comprehend fully, and it is responsible for what we see on the box. Within this camp a wide array of schools of thought (of the religious and philosophical kind) can be found, each claiming to know how the pieces fit together. The larger, more common of these schools call the creator-deity "God," while others call it the "Supreme Being," the "transcendent intelligence," the "prime mover," the "first causal agent," or something else. Aside from their mutual belief in "a" creator of some kind, many of these various schools have almost nothing in common. Some believe in the authority of a written text, while others believe in gnosis (divinely revealed knowledge). Some believe in an afterlife with a judgment, a heaven, and a hell. Others believe in one or two of these but not the others. Some believe in a savior, saints, and a Satan, but others believe in equal and opposite forces of good and evil. And because the overwhelming preponderance of these beliefs are speculative (not provable) and subjective (a matter of opinion), the variations seem endless.

With so many different possibilities, how could anyone ever hope to figure out how to put the pieces together to make the picture on the box? Frankly, nobody ever could . . . until, perhaps, now. This author has embraced the challenge and arguably figured it out, or at least fitted more pieces together than has ever been done before. If any parts are missing or improperly placed, it is hoped that others who follow will be able to build upon this work to complete the picture.

This puzzle analogy that is employed herein to introduce *The Explanation* has been useful, but it is imperfect. Why? Because the creation story told in the book of Genesis is not merely one of four equal corner pieces in the puzzle. It looms much larger than the others and is actually much more important than they are. It is the one with which all the rest must ultimately fit together if the picture on the box is to be made. Since it is part of the larger work called the Judeo-Christian Bible, understanding some basic information about the Bible as a whole is paramount to grasping how certain parts of the puzzle fit together. Thus, it should be helpful to make a few observations about the Bible now before doing anything else.

THE BIBLE

The Bible is a not a single book but a collection of books, divided into two sets of ancient scriptures, which are commonly called the Old and New Testaments. Although these texts were written mostly, if not exclusively, by people of Hebrew ancestry and Jewish religious background, the collection of them into what we call in English "the Bible" is a purely Christian concept, not a Jewish one. Christians often read the Old Testament as something akin to an appetizer for the main course, the New Testament. According to this mode of thought, the Old Testament is a set of writings that is, in a manner of speaking, inferior to the Gospels, the Acts, the Epistles, and the Revelation; it is foundational and provides prerequisite information about the only true God, but it does not show the path to salvation or eternal life. If we take a step back from this Christian conception, however, and look through the lens of an objective, inter-religious observer, we will see that the so-called Old Testament is really the story of when, how, and why God selected the Jewish people as his "chosen" ones, the ancient nation of Israel as his special country, and Jerusalem as his capital city or headquarters on earth. From this perspective, the fact that it contains passages that seem to presage Christianity is incidental. This in no way implies that those passages are not important; they are monumentally important to Christianity, but not so much to Judaism, which rejects the New Testament and sees the Old Testament (called the "Tanakh" in Hebrew and the "Masoretic Text" in English) as the main course (and some say "only" course).

The Bible is not a history book, although it contains a lot of history. It is not written in chronological order, although its modern, canonized format makes it appear somewhat so. Nor was it penned all at once like a monograph by a single author narrating a set of events, although many critical readers (mostly believers) find it amazingly consistent from beginning to end, as if it were written by one person or as if it were divinely inspired. Skeptics, of course, are quick to point out its inconsistencies and (apparent) contradictions which counter this divine inspiration argument. Regardless, there is so much that the Bible does *not* tell us about the past that we have no choice but to infer, extrapolate, and speculate in order to form any kind of cogent narrative from it.

The arrangement of the books of the Bible was determined by scholars living centuries after those books were written, and there is much disagreement today about what should constitute the canon of scripture—that is, what books should be included in the Bible. The question of whether the right books have been included and the wrong ones omitted has been up for debate for centuries. Likewise, what order these books should be placed in within the canon is a good question. Should the oldest book—the one most scholars say was written first—Job, for instance, actually come before Genesis, which scholars say was written (or better yet "finalized") hundreds of years later? Then the question of language and translation arises: should the books of the Bible be preserved in their original languages for purity, accuracy, and authoritativeness? Or should they be translated into a myriad of modern languages for convenient mass consumption and real-life application? If the latter, which interpretation of each should be used, who gets to do the interpreting, and how can we be sure of the accuracy?

The Bible can and should be questioned. If it cannot stand up to scrutiny, it should be discarded as a book of fairy tales. If it can withstand scrutiny, however, it should be considered at the very least to be "a" source of historical information and "a" source of spiritual enlightenment. Those who say it is more than just "a" good source—that it is actually the inerrant, infallible word of God—would do well to consider that, if so, it is definitely NOT the historical record of absolutely EVERYTHING there is to know about God or his will or his ways. In fact, it surely constitutes no more than a tiny fraction of what humans can know about God, which is why Jesus comforted his disciples at the Last Supper by telling them that He would soon send them the Holy Spirit which would guide them into "all" truth [John 16:13]. He did not say—and this is monumentally important—that the *scriptures* would guide them into all truth, but that the *Holy Spirit* would! The scriptures are rather for instruction, correction, knowledge, wisdom, direction, etc., but they are not, and indeed cannot be, the source of ALL truth. To think that the vastness of God could be contained in a mere collection of 66 books and a few hundred pages is ridiculous! If it were otherwise, there would be no need for preachers, teachers, and Christian authors to make sermons, lessons, and literature, for we would have nothing to say that wasn't already said in the Bible. As it is, however, we use the Bible as the main source of enlightenment to help us navigate through a world of darkness, we draw inspiration from it in hard times and direction in times of confusion, we make it the standard by which we judge right and wrong, and we even trust our eternal souls to the truth of its words.

With all that said, much fine knowledge and useful instruction for life can be found in all kinds of secular sources. We can, for instance, learn a lot from reading the history of ancient Greece and Rome as recorded by pagan scholars. Likewise, if we want to know how to build a mechanical device, such as a car, we don't look to the Bible, nor do we wait for the Holy Spirit to supernaturally reveal it to us. The Bible does not teach us mathematics, chemistry, physics, or most other fields of knowledge. That is why it is not, and cannot be, the source of ALL truth. If it does not tell us everything there is to know about comparatively trivial earthly things, how much more true must it be that it does not tell us everything there is to know about the infinite, eternal, ineffable one we call God!

It is a logical fallacy to assume that since the Holy Spirit leads us to the "rest of" the truth (all that is not contained in the Bible), then any seemingly revelatory thought that comes into our minds while we are praying or contemplating theology must be the voice of the Holy Spirit. The Bible clearly instructs us to "test the spirit(s)" [1 John 4:1], and common sense by any standard confirms the necessity of doing so. So how do we test the spirits? By recognizing and acknowledging that the Holy Spirit never leads us away from what has already been propounded in scripture but rather takes us into depths of understanding and interpreting the scripture that we could never experience otherwise. Skeptics and unbelievers will respond, "Aha! Gotcha! That is circular reasoning! You're saying the Holy Spirit serves as the voice of God, and the Holy Spirit says to believe the Bible, yet our only knowledge of this Holy Spirit character comes from the Bible. So that's just the Bible telling us to believe the Bible."

Of course, skeptics and unbelievers have other objections to the Bible being the word of God. They say the Bible is full of contradictions, and that proves that God is either confused, changes his mind on a whim, or else He is no God at all! They cite examples, starting with the fact that the God of law and judgment of the Old Testament looks like a quite different entity from the God of grace and forgiveness of the New Testament; and the concept of an afterlife with a heaven, hell, judgment, rewards, a purely evil "Satan" character, etc., *mostly* seems to have been absent from the Hebrew/Jewish religion and only added over time by certain Jewish sects that ended up creating or embracing Christianity. Does this not prove beyond a reasonable doubt that humans made God in their own image, they rhetorically ask, rather than the other way around? For many seekers of the truth, confronting this line of reasoning for the first time stops them in their tracks and puts an abrupt end to their quest. But people of faith like myself, of course, answer that this is proof of nothing other than the fact that we humans are limited in our understanding of the mysteries of God, and we must thus continue the quest.

Some believers will continue the quest by avoiding such questions, others by giving superficial answers and then moving on, and still others—Christian apologists—by confronting the challenge of answering the hard questions posed by non-believers. *The Explanation* falls squarely within that latter category. Answering each and every specific question and responding to every individual charge that skeptics make about the veracity of the Bible, is obviously impossible in this short book. For now, though, we can answer the question posed above pertaining to whether God is made in man's image rather than the other way around. The answer is that God has not changed, but our human understanding of God has changed greatly over the centuries, from generation-to-generation, nation-to-nation, and language-to-language. As the apostle Paul put it, "We see through a glass darkly" for now, but someday we will see clearly [I Corinthians 13:12]. Although skeptics and doubters may take this as a cop-out, it is not. We are not saying, "Since we can't possibly understand God today, why try? Let's just give up and live by blind faith." That truly would be a cop-out. Instead, we are saying, "We don't see the whole truth yet, but we are inching closer and closer to it all the time. Let us continue to make the effort; let each generation build upon the work of those who came before."

We can also answer another similar line of questioning right up front: if the Bible does not contain "all truth" but only points us in the right direction until the Holy Spirit takes us the rest of the way, how can we know when we are hearing the voice of the Holy Spirit and when we are hearing our own consciences or some unholy spirit? How are we to know whether a thought that pops into our minds is a divine revelation or not—is it when it comes while we are asleep and is unusually vivid and realistic? Or is it when it comes while we are wide awake and gives us goose bumps or otherwise causes an intense, involuntary, emotional reaction in us? Or maybe when it occurs under hypnosis or puts us into a trance once it begins? Actually, none of the above. All of these possibilities are examples of mysticism. God is mysterious to us earthlings, of course, but He is not mystical in his dealings with humans. He has done, and probably continues to do, many things that appear to humans to be mystical (translate as: "it seems like magic") because our limited understanding prohibits us from figuring out how it happened or how it is possible. But that is different than saying "there is no explanation for it." Even miracles have an explanation; we just haven't found the physical explanation for them . . . yet.

Consider this: most specific cases mentioned in the Bible where God deals one-on-one with a person to give him revelatory knowledge are physical in nature. Starting in Genesis, when God "walked" in the Garden of Eden in the "cool of the day" [Genesis 3:8] and talked to Adam and Eve, the description is that of a human-like being in the flesh making actual, audible conversation, not a ghost or an invisible voice coming out of an otherworldly ether somewhere. Continuing all through the Old Testament and into the New, this is the pattern. In the book of Acts, for example, God struck Saul down and blinded him on the road to Damascus [Acts 22:11]. While others around Saul heard no voice, they all saw the light which physically incapacitated him, rendering him literally blind and helpless. Even in cases in which it seems at first glance that God spoke to some biblical patriarch through a mystical vision, there was physical evidence before, during, or after the fact to confirm it. Example: Jacob wrestling with an angel in his sleep [Genesis 35: 1-7]. It was not merely a dream, for when he awoke, he bore the scars of an actual battle. When Moses went up on the mountain to confer with God for long stretches of time, others around him naturally doubted the legitimacy of his claim to having a face-to-face meeting with God; they assumed that he was spending all that time on the mountain making up a list of rules for them to live by in order to control them for his own selfish purposes. His words were confirmed, however, with miracles and superhuman powers.

It would, of course, require a whole separate study to analyze every case in the Bible where something that appears mystical happened and try to explain how it occurred—God speaking through a donkey, God making the sun to stand still for a day, Jesus walking on water, Jesus turning water into wine, etc. Such a book would be an outstanding supplement to the present volume, and perhaps it can be written in due time, but for now let us stay focused on the subject at hand—explaining the first few chapters of Genesis in light of Christian theology, secular science, pagan history and mythology, and alternate theories.

Of all the hard-to-explain things in the Bible, perhaps none is more difficult than the Trinity. Yet, reconciling the Christian doctrine of the Trinity (or "Godhead") with Old Testament monotheism, and particularly with Genesis, is absolutely vital to formulating a plausible rationale for why we should take the Bible as a whole seriously. The concept of the Trinity has twisted some of the greatest minds of the past 2,000 years in knots. Attempts at theological explanations for it abound, but most believers ultimately just end up taking it on faith as a fact without understanding it. For it to be plausible to unbelievers and skeptics, however, it must have a *logical* rather than merely a *theo-logical* explanation. It must be based on rational thought rather than faith, in other words. Rational thought (also called "reason") is a step-by-step process, not an all-at-once leap to a conclusion. *The Explanation* will thus lead the reader through those steps. A good deal of patience, persistence, and focus will be necessary on the part of the reader to follow the steps and connect the dots; it would be easy to lose sight of the proverbial forest herein as we begin analyzing all that can be analyzed about any given tree.

Another vital point that must be made right up front is that Christians and other monotheists tend to think of God as a masculine being and refer to "him" accordingly, because the Bible uses that terminology. This way of trying to understand and describe God is imperfect yet useful, as we shall see, because it allows us to visualize and discuss what would otherwise be unfathomable and indescribable. It gives us a starting point, in other words, for communicating on the subject of God. So, this choice of words is used very deliberately in *The Explanation*, but the caveat is that using this masculine pronoun does not *limit* God to the form of a human male (with emphasis on *limit*). God is not, and indeed cannot be, limited in such a way; that much is certain. But there is so much more that we can't know, so we must make assumptions about the rest. In order to construct a plausible overall explanation for the existence of all things, each assumption we make about God must be reasonable in its own right, and each must lead to the next in some kind of logical sequence. That is precisely what *The Explanation* sets out to do—follow a series of logical points that interconnect, converge, and ultimately funnel toward a plausible conclusion.

Finally, we must concede the fact that each theological question we ask leads to yet more questions, and it seems that we could go on asking questions *ad infinitum*. Such is the nature of theology, but especially of ontology (the study of the first cause or prime mover of creation). Indeed, asking questions long enough and thoroughly enough will inevitably bring us to classics of theological abstractionism such as, "How many angels can dance on the head of a pin?" And "Can God make a boulder so big that He can't move it?" Yet, the purpose of *The Explanation* is, inasmuch as possible, to answer questions, not continually ask them. We must also acknowledge that what passes for the "orthodox" (right, proper, correct, or widely agreed upon) answer to any of these questions is almost completely subjective, so no honest, humble person should claim to have "figured it all out" while judging that everyone else has it wrong. With that said, it is incumbent upon all literate believers to try to grow in understanding of the Bible, just as it goes without saying that all rational humans should, regardless of their religious beliefs, try to grow in scientific knowledge about the world we live in and about the cosmos in general. Learning and knowing begin by asking, probing, speculating, theorizing, doubting, countering, rejecting or accepting, and ultimately synthesizing. In the end, science, philosophy, parapsychology, and theology all must match up, or come as close to matching up as the human mind can make them, in order to formulate a credible explanation of what really happened "In the beginning."

While using Genesis 1:1 and whatever came before it as a point of departure for our intellectual journey (which we can visualize as a train ride), we will pause along the way at various biblical junctions that force us to veer this way or that; we will stop at scientific depots that make us get off and either change directions or change our mode of transportation entirely; we will park at philosophical inns that invite us to spend the night resting and thinking; and we will pass by miscellaneous landmarks such as conspiracy theories and historical facts that cause us to wonder and ponder before moving on. At the terminus of our journey, we hope to have arrived at our intended destination—the truth. The truth about exactly what, you ask? The truth about whether the first few chapters of Genesis should be taken seriously as a source of information on how the universe came into being, where all that exists (especially living things) came from, who made it all or at least put it here, whether there was/is a plan and/or purpose to it all, and if so what the plan/purpose is.

Prequel: Before The Beginning

The Bible opens with the words, "In the beginning, God created the heavens and the earth." This opening statement is teleological (indicating a purposeful and orderly creation), and it makes a perfectly logical starting point for the Bible, if we presume that the Bible is intended as a sort of road map or set of instructions to show humans how we can know God. It makes for a disappointing and frustrating start, however, if we presuppose that the Bible is intended as an explanation of the origin and development of all things. It immediately raises the seemingly unanswerable ontological question in the mind of thoughtful readers: "Where did God come from?" or "Who created God?" In order to plumb the depths of Genesis 1:1, make sense of all that follows from it, and inasmuch as possible reconcile it with secular scientific knowledge and rational thought, we must discuss this prerequisite topic of what came *before* the "beginning" mentioned in the first line of Genesis.

At first glance it appears that the Bible does not address what came before Genesis 1:1, but actually it does, although this information can only be gleaned in bits and pieces scattered throughout its various books. By taking those biblical fragments and combining them with information extracted from other ancient texts, the archaeological, geological, astronomical, ecological, and biological record, then applying some critical thinking, we can produce a systematic theology and/or complete worldview that validates Christianity as the best (if not proves it to be the only true and right) religion. Achieving this lofty goal will be painstaking to say the least, but only by doing this hard work can we hope to understand what has always before seemed impossibly mysterious.

Christ's disciple John left us a testimony of the precarnate identity of Christ, calling him "The Word," equating him with the creator of Genesis Chapter 1, and saying He was "with God" while simultaneously *being* "God" [John 1:1-5]. This can be as confusing as anything one might ever encounter theologically, because it states a non-earthly, non-human truth by using earthly, human language which is incapable of adequately conveying that truth. John was writing a Gospel, not a commentary on a Gospel, however, and He knew that the truth of his succinct statement would be grasped by some (those who were chosen of God to receive it, which Jesus called "the pure in heart" [Matthew 5:8], the apostles called "the Elect" [I Peter 1:2], and which we might simply call "believers"). So the Holy Spirit led John to write it in this cryptic form for the same reason Jesus spoke in parables—to prevent the impure in heart from grasping it. One of the main goals of *The Explanation* is to be that commentary which John's Gospel is not, and thus to explain the mystery of existence. This explanation is possible only if we board a logical train of thought, which starts as follows.

THE INDESCRIBABLE ONE

At the beginning of what we humans know as "time," the supreme being that we call "God" created the physical universe and everything in it, including time itself. So God existed before time existed. Time was and is merely part of the physical creation, as is all matter (such as the elements and all they combine to make) and properties (including space, motion, light, sound, speed, strength, color, weight, density, gravity), and whatsoever is observable or quantifiable within the universe. As the creator of time, God exists outside and apart from time, and is thus not limited by the arrow of time as are all physical beings and things. God therefore experiences the past, present, and future not as discrete divisions of time that unfold in a sequence, but instead as three simultaneous occurrences. Time travel is therefore possible but unnecessary for God, because He is already "there." That is how He knows the end from the beginning, and that is why, taking the form Yahweh ("the Lord"), He identified himself to Moses by a name that signifies the ever-present state of being "I am" [Exodus 3:14]. Some scholars believe a better translation is "I will be what I will be." Alternatives to that translation, which give the same meaning in different words, are "What I was and am is what I will be" and "I will be whatever I want to be." Perhaps the best alternative, however, is actually the simplest one: "I exist."

For Christians this concept of the ever-present state of being explains why the resurrected Christ could rightfully call himself the Alpha and the Omega, the beginning and the end, in Revelation 1:8, 21:6, and 22:13. Despite coming to earth in the form of an embryo-turned-infant-then-child-and-man, He had already existed in a precarnate form as one manifestation of God—the one called Yahweh in the Old Testament. That is why when the Roman soldiers came to arrest Jesus in the Garden of Gethsemane on the night before his crucifixion and they asked him if He was the one they were looking for, He answered, "I am," instead of answering "Yes" or something else. When He spoke the words "I am," power emanated from his lips such that the soldiers fell backward [John 18:6]. Clearly the Lord was restraining himself when He answered them, knowing full well that He could have uttered those words with the same force and impact they had when He spoke to Moses from the burning bush. If He had not restrained himself, He would have given away his identity such that his accusers would not have crucified him. He knew that dying as the atoning sacrifice for the sins of all humans was what He was sent to earth by his Father (the "Elohim") to do [John 18:11], so He could not reveal his identity fully. Yet striking the perfect balance, He gave his accusers a hint that would later serve as a witness against them by uttering the words "I am."

The creator-God was and is first and foremost a "spirit" (a living consciousness or mind). This spirit-God is called the "Elohim" in Hebrew, and this is the specific name for God that the narrator of Genesis uses in Chapter 1. The Elohim can manifest in a tangible form inside the physical universe when choosing to do so, as when becoming Yahweh Elohim, "the Lord God," in the Old Testament, and Jesus "the son of God" in the New, but when He does so it is not an either-or, all-or-nothing change. He simultaneously remains a spirit existing beyond time and space. He is at once the Elohim and Yahweh, the Elohim and Christ, the Father and the Son, a spirit and a physical being; not two Gods, but one God in two forms. So when Jesus taught his disciples to pray by saying, "Our Father which art in heaven . . .", He was referring them to the non-physical, time-less, eternal Elohim, of which He was himself one manifestation. He did not refer them to his precarnate self Yahweh, which was just one manifestation of the Elohim, for that would have been tantamount to his telling them to pray to himself! Nor did He refer them to the Holy Spirit which was another manifestation of the Elohim, and which they would not have understood anyway, because He had not yet taught them about the Holy Spirit. This all explains why the name "Elohim" in Hebrew is written in the plural as if referring not to "*a* god" but to "*the* gods," yet is always placed in a context in the Bible that emphasizes the unity or oneness of God. We never get the sense that there is more than one God or that these various manifestations of God function like self-contained or self-willed individuals who merely work together as a team for a common purpose. Thus, the Elohim is one God that can manifest in any form He chooses, not just in the three that Jesus mentioned specifically as "Father," "Son," and "Holy Ghost." (Don't misunderstand: this does not mean Jesus got it wrong! It means our orthodox understanding of Jesus' words is wrong.) That is why in *The Explanation* the word "Elohim" will be prefaced by "the," indicating neither an individual nor a group but rather an *entity* which is indefinable using conventional language, but might best be conceptualized as something infinitely flexible.

It is this entity which the apostle Paul called the "Godhead" [Colossians 2:9] and which theologians have tried, mostly in futility, to explain using the concept of the Trinity—a term which does not appear in the Bible but in theory means the same thing as Godhead. One reason the concept of the Trinity is so hard to understand is that it requires us to visualize God in three "persons" (Father, Son, and Holy Spirit) instead of two "forms" (spirit and physical body), yet two of those "persons" (Father and Holy Spirit) are of the same "form" (spirit). This is obviously confusing because it contains an unnecessary redundancy. By training, habit, and desire to maintain orthodoxy, many, if not most, theologians are stuck on the idea of the triune nature of God and thus will not consider or take seriously the possibility of the concept of the Elohim as a duality of forms. Jewish theologians are locked in to thinking in terms of a mono-God, and Christian theologians are locked in to thinking in terms of a triune God. So collectively, they have decided there can be only two possibilities for who and what God is—He is one, or He is three. Yet, everything in the Bible points to God in terms of manifesting in just two forms—spirit and body. When we conceptualize God as a duality, we actually see the Father (a spirit that must be worshiped "in spirit and in truth"), the Son (which needless to say has a body), and the Holy Ghost (which is obviously also a spirit). If indeed there are two spirit manifestations of the Elohim as suggested here, what is the difference between them? Why should there be a "Father" and a "Holy Ghost" if they are both spirit? The answer is that the Father never enters the creation (the physical universe) but exists apart from it always, whereas, like the Son (Yahweh in the Old Testament and Jesus in the New), the Holy Ghost is dispatched by the Elohim into the creation. That is, the Holy Ghost is the part of the Elohim whose purpose is to deal specifically with humans in the physical realm.

God can and does step inside the physical universe sometimes in bodily form, as evidenced by Jesus' life on earth as the son of Mary, and, as we will see in the next few chapters of *The Explanation* in our discussion of Yahweh's physical presence in the first four chapters of Genesis. When He steps inside his creation, however, He is not confined within it, as all created things and beings are. Although visiting earth on occasion and manifesting in the presence of humans in bodily form, whether as the Old Testament "Lord" or as the New Testament "Son," God is infinitely greater than the body He inhabits and can manifest inside it and remain outside of it at the same instant. When God manifests in a physical form, in other words, it is not an exclusive change; He remains the Elohim. God does not have to be either/or, but can be both spirit and flesh simultaneously, both here and there at the same time. In that sense, the monotheistic religions (Judaism, Christianity, and Islam) which conceptualize God as either an individual or a Trinity are both missing the mark by a degree or two, while the non-monotheistic religions (various pagan and mystical religions) which conceptualize God as a multiplicity or amorphous mass are actually closer to hitting the mark, or at least no farther off. The Holy Ghost is not, as He is so often described, the "third *person*" of the Trinity. Instead, He is one manifestation of the Elohim, just as the Lord (Yahweh) of the Old Testament is one manifestation and the Son (Jesus) in the New Testament is one manifestation. These manifestations are not separate and discrete beings; they are for lack of a better way to put it "part" of the Elohim, having the same essence—thinking as one, speaking as one, and acting as one. They cannot disagree because they *are* one. So to recap, God is a "time"-less being called the Elohim who can move effortlessly into and out of any state of being, manifesting as spirit and body—two forms, not one or three—simultaneously.

God is, of course, the supreme intelligence, and all created beings, including humans, are necessarily inferior and always will be. We will never reach a point where we know as much as God and can do as much as God or be what God is; no matter how much we progress as a species—even if we master time travel—we can never ascend to the level of equality with God. The original sin which the created being called Lucifer in the Old Testament committed was thinking that He could. Having existed for eons before the creation of humans, Lucifer had progressed in his knowledge until He thought it possible to be equal to or greater than his creator [Isaiah 14:12-14]. He clearly had capabilities that are far beyond our comprehension, which is why his temptation to be God's equal makes no sense to us when we read about it in the Bible. As of now, we humans have not progressed far enough in our knowledge to think so highly of ourselves, although non-believers, in pushing the frontiers of scientific knowledge daily, are unwittingly driving the species in that direction, thus falling into the same temptation that brought Lucifer into eternal damnation as the being that became known as Satan. Since they think there is no God, and since they have as yet found no life form in the universe superior to themselves, they become their own collective god, which is an alternative version of the same sin that Lucifer, who knew very well of his creator, committed.

Being so much greater than us, God is not, and indeed cannot be, easily understandable to us, for we are but mere finite products of his infiniteness. At best, we can hope to understand only those aspects of his being that He chooses to share with us. God is likewise not easily describable to humans, because we are limited by our need to visualize everything in terms of our familiar earth-bound existence. We are forced to use earthly comparisons, to visualize God as being like this or like that—things to which we and our fellow humans can relate from experience, observation, or education. Those of us who have been brought up seeing pictures of God looking like a man with a long white beard sitting on a throne in the clouds tend to visualize God that way into adulthood, if not all the way to the grave. We are prone to substitute such mental images for the truth, and the truth is that none of us knows what God really looks like, although some humans had the luxury of seeing Christ in the flesh and thereby beholding one of his manifestations. (And if the Shroud of Turin is the authentic burial cloth of Jesus, we too now know what He looked like, thanks to a History channel television documentary called "The Real Face of Jesus." Even if it is not authentic, the three-dimensional image that has been produced from it still gives us a good idea of what He realistically *might* have looked like, and that is better than purely guessing at it as artists have done for centuries.)

Although humans were made, according to Genesis 1:27, in the "image" of God, this image is not the direct reflection of the fullness of God. It is rather merely *a* reflection of *one* manifestation of God, because God cannot be confined to one form or appearance. It is much like you or I looking in a mirror. We see an image of ourselves, but it is only two-dimensional. It does not capture our fullness in three dimensions. It can show us only one side of ourselves at a time. If we therefore assume that "image" means God has one head, two arms, two legs, two eyes, etc., like humans, we are projecting our own physical form onto him. In so doing, we are guilty of trying to make God conform to *our* image. Or, put another way, we are guilty of trying to create a God in our own image. That is, of course, exactly backward from the true meaning of humans being created in his image. So to repeat, God is a being too great to be confined to a single manifestation, an unchangeable shape, or a limited body.

Humans were also made in the "likeness" of God, which, at first glance, appears to be a redundancy. Yet the terms "image" and "likeness" as used in Genesis are not synonymous. "Likeness" does not refer to physical appearance. Instead it refers to having certain internal characteristics, such as a high-functioning brain that is capable of feeling and displaying a wide range of emotions, thinking in abstractions, knowing right from wrong, possessing intuition, and having an eternal spirit or consciousness despite being trapped in a mortal body. Although we have his likeness, it is and can only be a partial likeness. Just as a raindrop has the likeness of the ocean, so are we tiny things made with some of the same substance as the infinite God. The purpose of pointing out these obvious characteristics of God is to show from the outset that, despite the fact that we are created in his image and likeness, there is only so much that any human can possibly know, or to even have intelligence enough to speculate about, for that matter.

For reasons known only to himself, at some point in the time-less past of eternity, God decided to create physical things. Some people have speculated that God became a creator because He was lonely, and others have rightfully asked how that could be possible. Some have speculated that God felt a need for adoration—to be worshiped—but others have again asked how the supreme being could be lacking something He did not already possess. Yet others have postulated that perhaps God wanted someone or something to entertain himself with, to occupy his time. There have been other theories put forth which hold, for instance, that God wanted a "terrarium" or "ant farm," so to speak—a creation that moved of its own accord that He could observe as a scientist might observe insects inside a glass case. These ants would have free will to build any world they chose, so long as they remained inside the space allotted to them and built with the materials available to them. God would simply "watch" to see what choices they would make or what kind of world they would build. He would not interfere with it, for his plan was neither to "steer" it in any particular direction nor to "fix" it if it went astray, but simply to observe it out of curiosity. A similar theory is that the creator acted as a "clockmaker" who built a machine (the universe) and ordered it to function according to natural, mechanical laws, so that He would not have to do anything with it, to it, or about it other than observe it. He would be free to intervene in it, or interfere with it, on rare occasions, but would generally not do so. Cases in which God manifested within the machine He had built would appear from the vantage point of humans to be "miracles"—moments when God pauses the natural laws by which the machine otherwise runs in order to adjust something in the machine or reveal himself to the parts of the machine made in his image.

All such theories must be judged wanting, however, because if God had been "in need" of anything, He had the power to will into existence the fulfillment of that need immediately, such that never did He do without or thus "suffer" for even a moment. Indeed, God could not and cannot be deprived. He always was, still is, and always must be, complete in himself—having no wants, no needs, no fears, no failures, and no possibility of ever lacking anything. He is, in short, "perfect." Here our human understanding breaks down; here our ability to visualize, describe, and explain fails us, because we cannot fathom perfection. Although we try, we deceive ourselves if we think we can. On earth, all we have are rough approximations of perfection, because everything here is temporary (fixed in time), and thus imperfect. If a thing cannot exist forever, it is not perfect. At best it may be considered "temporarily perfect," and even if we concede that such a thing might be possible, it falls short of the ultimate perfection—eternal, infinite perfection. We may try to conceive of perfection in any number of ways using our earthly mind. For example, we can imagine (or experience) a seemingly "perfect" day, with weather that is just right—not too hot, not too cold—in which we feel a seemingly perfect love for a significant other who we are sure is also perfect. On that day, we are in perfect health—in the prime of life, at the peak of our mental acuity and physical strength. We are strolling together in perfect harmony, carefree, enjoying the moment to its fullest, feeling as though it will last forever, knowing true happiness, peace, and contentment. In due time, of course, that day passes, and soon we discover that the object of our "perfect" romance is anything but perfect. He or she is just as imperfect as all other mortals. Time is the villain in this analogy. Perfection eludes us humans because we cannot stay in the moment but are compelled in a linear, forward direction by time.

Perhaps a better way of conceiving perfection is to visualize a seemingly perfect diamond—flawless to the eye, even under a microscope, having no discernible faults. Left to itself, it may exist in that state for eons and eons—and we might even say for "eternity," except we realize the concept of eternity, like perfection, is something humans cannot quite grasp. Time, therefore, does not seem to be the enemy of the diamond as it is to our "perfect" night of romance. As perfect as the diamond may seem, however, it is still imperfect, because it is an inanimate object: it cannot move of its own accord, it does not have a mind, and it does not have a soul. God is not limited by either time or lifelessness, of course, and thus truly is perfect. And yet, again we are aware that He chose to become a creator, and we concede that we do not know why, but we can be certain that it was not because He lacked for something. The best explanation for why a perfect, supreme, complete, spirit being would become the creator of a physical universe is that He had a plan which we are not allowed to be privy to until it is finished. Then it will be revealed at some point in either the future (as linear time goes) or in another realm (such as Heaven or another dimension) to all of the creation, but mainly to the special creation made in God's image—humans.

THE LOVE STORY

Although it is impossible to ascribe a sequence to events that happened before the creation of time, for purposes of our limited, earthly understanding, we can say that God's "first" physical creation (for lack of a better way to put it) was an individual bodily manifestation of himself. The physical version and spiritual version were and are the same being. They share the exact same mind and consciousness; they must necessarily agree at all times on all points of decision, therefore; they cannot disagree. The spirit God can be considered as the "Father" and the physical God as the "Son," to use standard Trinitarian Christian terminology. Only insofar as the spirit-God preceded (figuratively speaking) the physical God in the time-less past, the Father is greater than the Son. (We only say "preceded" because we cannot otherwise describe an "event" that occurs in a state of timelessness.) According to John 4:24, the Son, while on earth in the form of Jesus, referred to the Father and himself as one, yet deferred to the Father as being greater than himself, which on the surface seems to be a contradiction. The explanation is that the two are joined, but not in the sense that, for instance, a husband and wife are joined or two separate businessmen come together to form a partnership. These earthly analogies are in fact exactly backward from how the spirit Father and the bodily Son are joined, because they require beings who were first separate and distinct to find each other and agree to interact. In the "heavenly" or cosmological realm, the Father and the Son started off together as one, and the Son proceeded forth "from" the Father. In that sense, the Son was never technically "created," but a bodily form of God simply proceeded from the spirit form.

It is important that the reader understand that this was not the event in which the Son was "begotten" of the Father, because the term "begotten" in the Bible refers to the process of sexual reproduction. We know, of course, that at a particular point in earth's "history" (the time that is measurable to humans and in which events are recorded and dated by people), Jesus would be physically "begotten" when a female homo sapiens of Jewish lineage named Mary gave birth to him in human form. Yet, during eternity before time was created, the Son proceeded forth from the Father in a way that we might best visualize (to use another imperfect analogy) as a drop of water falling from a cloud. Here, both droplet and cloud are the same substance—H_2O—just different manifestations of it; the droplet is never anything other than the same substance as the cloud from which it proceeded despite its change in form, and it can always be turned back into vapor to become part of the cloud once again. During the time the droplet is separated from the cloud, the cloud does not cease to be a cloud. It does not lose its "cloudiness" just because a part of it has temporarily been removed from it. Likewise, the droplet never ceases to be H2O; it can exist apart from the cloud and still be just as much the same substance as the cloud itself. Besides this analogy of the water, perhaps the closest earthly equivalent that we might find to this process is that of a single-cell organism dividing into another and thus replicating itself. These earthly analogies break down, of course, in that the Father and the Son did not separate into distinct beings that exist *exclusively* apart from one another, but remain both together and separate at the same time. Again, our human minds cannot quite grasp how this is possible, because indeed it is not possible in the earthly, physical realm in which we are trapped. This limitation notwithstanding, we press on trying to understand as best we can.

Therefore, consider this: when Jesus walked the earth in human form some 2,000 years ago, He proclaimed in John 4:24 that his Father was a "spirit," adding that if humans want to worship him, we must worship him in "spirit" and in "truth." Exactly what this means is a mystery, but one must suppose that it means, at least in part, that worship must include a combination of emotion (specifically the feeling of love, longing for, or attachment to him) and intellect (specifically an attempt to discover, understand, and be in good standing with him). Rival church groups—what we commonly call "denominations"— emphasize one or the other of these and differ over the amount of each that is necessary for worship to occur as God intended it. Like the birth of Jesus in human form, this is a whole separate topic which we will return to later. For now, let us consider the differentiation between the Father and the Son as revealed in the Christian

Bible. At one point as Jesus walked the earth in human form, we see in the scriptures that a mysterious voice was heard from above saying, "This is my beloved son in whom I am well pleased" (Matthew 3:17). Clearly, the voice did not come from the mouth of the fleshly Jesus but came from some unseen source in "heaven" (meaning the sky or some invisible dimension), if indeed this Bible story is true. It was reputed and understood by those present to be the voice of God the Father, although we might well wonder how a spirit can utter audible words in the physical world. Clearly, if we believe this passage in the Bible to be true, the spirit-God could and can manifest in the physical world in two different forms at the same time—in the form of the fleshly Jesus and in the form of an audible voice. This shows again that God is not limited by any earthly, physical constraints, and it reminds us how difficult it is for humans to grasp the concept of someone or something being in two places at once. Yet quantum physicists assure us that it is possible for certain subatomic particles either to be or at least appear to be in two places at once, so if secular science says it is possible in the physical realm, how much more possible must it be in the unseen spirit realm!

According to John Chapter 1, Jesus existed "in the beginning" in a form called "the Word," and it was He who created all that exists in our physical realm. But it was not the begotten Jesus who did it, but rather the unbegotten Son who proceeded forth from the Father in the time-less past. This may seem like splitting hairs in order to arrive at a distinction without a difference, but it will become vitally important shortly. Although we have just seen that the spirit-God is completely capable of manifesting in the physical world as a voice, when in Genesis we read repeatedly that "God said. . .

,") we understand it to mean that his bodily form spoke the words that brought all other physical things into being: time, light, the atmosphere, the dry ground, the oceans, etc. Why it would be Jesus who did the actual creating rather than the Father is explained by another biblical passage. The same apostle John, who begins the Gospel account that bears his name by calling Jesus "the Word," later generically calls God "love" (1 John 4:8). Although yet another mystery, we can make some observations about this terminology and arrive at some understanding of what it means, while being careful not to reduce it to something more elementary than it really is. First, our human minds automatically want to take the statement "God is love" and jump to the conclusion that the reverse must be equally true: "love is God." But that is not only incorrect, it is also a major, blatant fallacy of reasoning. There are at least four different kinds of "love" as defined by the ancient Greeks, and which of these, if any, the apostle was referring exclusively to, is debatable. But to assume that God would be limited to any of them, or even all of them, is to misunderstand the unfathomable greatness of the creator.

It may be, in fact, that "love" was the closest approximation that the author "John" could conceive of in trying to capture the essence of God, but it fails us as an explanation of the real point he was trying to make. Perhaps a better word would be "perfection." Consider it this way: if we rendered a different, better English translation of the point he was making rather than performing an actual Greek-to English word-for-word replacement, we might get "God is perfection" rather than "God is love," for the reasons previously cited herein in our discussion of the word "perfect." The concept of "perfection" includes, but is not limited to, absolute love. Purists may well take issue with this alternative rendering of the scripture into English, but if so, they must contend with the complexities of disproving the heretical thinking that would say since God is love, then love is God; *ergo*, God is not a "being" but rather a "feeling" or "emotion" or merely an abstraction representing all that is good. That interpretation of the scriptures logically leads to the notion that Jesus was born of "love" and/or became the "son of God" by being a "child of love," or even worse, a "love child" of Mary and (ostensibly) Joseph. Moreover, it can lead to believing that Jesus was the "son of God" only because He made the ultimate expression of love for his fellow humans by sacrificing himself on the cross for our sins, and in so doing basically taught us how to love in a more Godly (superhuman) way. Therefore, He showed us "God" by showing us a more intense, meaningful form of "love" than we would ever know otherwise. He pointed us to his "Father" by pointing us to a higher love.

To that train of thought, orthodox theists would respond, and rightly so, that God must be something more than that—more than a mere abstraction that changes our behavior for the better. Surely, that cannot be what John meant when he said "God is love." No, because "love" cannot speak from the sky and say, "This is my beloved son, in whom I am well pleased," yet the Bible clearly states that to have happened. (Likewise, when I John 1:5 says, "God is light," we know to take that verse allegorically, because light cannot speak from the sky with an audible voice in a human language. To say that God is literally "light" is to reduce him to a type of physical energy or an observable thing within the universe, but we know He is the creator of light, not merely light itself.) God is thus much more than our earthly understanding of what love is. If we were to say that love is the defining characteristic of God, however, that might be accurate. If so, it could explain why Jesus, the son of God, put into motion the systematic creation of this whole separate physical universe: to have a place within which to craft his Father's "love story," a drama in which He would play the starring role.

The plan of the fall and redemption of man was scripted before the first human was created. The script has never been seen in totality by any living human, but a few have seen fragments of it, given through revelation. These fragments, pieced together, mostly form the Christian Bible, although what "books" should constitute the Bible, how translations are made, and what interpretations are correct are, needless to say, subjects of much disagreement by theologians, lay people, and non-believers alike. We need not concern ourselves with those points of disagreement here, for we want to focus instead on the bigger picture—the love story. Parts of the love story are not found in the Bible at all. These parts must be inferred from other sources. Apocryphal writings, pre-Christian and non-Jewish history, archaeological evidence, scientific proofs, and even myths and conspiracy theories must be taken into consideration. When done properly, these extra-Christian sources help complete the fragments given in the Bible, and help verify (certainly not disprove) the basics of the Christian love story.

THE CELESTIAL DRAMA

With this background established, God's second (figuratively speaking again, for lack of a better way to describe a time-less event) physical creation was an other-worldly abode that we commonly call "Heaven," which is an actual, geographic place where, presumably, the physical son of God has resided for his whole existence except the few short years

He spent on earth in the form of Yahweh in the Old Testament and Jesus in the New. In fact, although we call it here the second object of creation, it seems that it could have been created simultaneous to the physical son, because how could a physical body with humanoid features survive in a non-physical environment? But again, we must not put limitations on God by rationalizing that since we humans can't do something, then Yahweh/Jesus couldn't do it either. Even so, for us to hope to make sense out of this mystery, we must start with certain assumptions. If indeed the son had a physical body in a place called "Heaven" before He arrived on earth as the baby of Mary, it seems logical that He lived in a place with air, soil, water, etc. This seems so obvious that we might well even make the leap that surely the Father created this physical environment for his son before He created the son. Yet the scriptures clearly state that everything was made by the son. So we must take that at face value, realizing that time did not exist before the creation of the universe, so saying which came first—the son or Heaven—is very much a chicken and egg riddle which leads to nothing more than a set of circular questions and answers.

So God created Heaven, but where is it, and what can we know about it presently? We should point out the possibilities that may or may not be true about it first. Some common theories are that it exists A) outside our physical universe, B) in another dimension of this universe which humans cannot perceive, C) within our observable universe but is either so far from earth as to be undetectable by humans, or is D) deliberately hidden from human detection by some divine space cloak. We cannot be certain which if any of these is the case, so let us explore each option. In the first case, Heaven might be somewhere beyond the physical universe. It might in fact be anything and everything that lies beyond the physical universe. If so, humans are not capable of visualizing it. After all, what would something look like that is not bound within the observable universe? Would it contain the same spectrum of colors, or would there be colors that don't exist in our known universe? Would it contain the same shapes, or would there be shapes that we can't imagine? What laws would govern it, if not the same laws that govern this universe? Would, for instance, gravity exist there? These are, of course, just a few examples of unanswerable questions stemming from this option.

In the second case—another dimension—this observable universe and Heaven, along with all it contains, exist side-by-side and simultaneously. In this scenario, beings in the heavenly dimension have the ability to move from their realm into ours, but we do not normally have the ability to move from ours into theirs. The time when humans can and must move to the other dimension is in death, but also in cases of the miraculous, when God allows them to be transported there temporarily in order to give them a message to take back and share with fellow humans in their own dimension. If this theory is true, humans are basically at a God-ordained disadvantage that we, presumably, cannot overcome. That is, it would be natural to assume that if God wanted us to be able to move into the heavenly dimension in this life, He would not have set up the system the way He did. However, it actually may be that He wants us to try all the same, and it may be that in time scientists can figure out the "technology" (for lack of a better word) to make it happen. For it may be no different or harder, relatively speaking, than earlier humans discovering that the earth is round and is a sphere floating in space in orbit around a sun in a small system on the outskirts of a galaxy, etc. In other words, given sufficient time and research, it might be doable, and God may allow it. This is not as implausible as it seems at first glance. God may indeed have set up the whole process just for that purpose—for humans (the one creation made in his image)—to be able to work his way back to the creator through his own (scientific) efforts. Before dogmatic Christians immediately dismiss this theory, they would be wise to consider that it is not necessarily or automatically in opposition to the redemption story of Christ. In fact, it actually complements the love story, as will be shown later in *The Explanation*.

In the third case, much the same kind of logic applies. If Heaven is an actual, physical place within this universe, but is simply so far from earth as to be undetectable at present, it stands to reason that in time human ingenuity should progress to the point where we eventually find it and figure out how to get there. Whether through some type of faster-than-the-speed-of-light spaceship, a time machine, a worm hole, or whatever, science should be able to conquer the distance, although we cannot know how long into the future we would have to wait for it. In the fourth case, if God deliberately hid his abode from humans, then it is not possible under any circumstances for humans to find it; it will remain hidden until such time as God decides to unveil it. So what is the explanation? Actually there is a better explanation than any of the four above. Heaven is actually a combination of physical and "spirit" world, where physical properties exist and apply–linear time, the speed of light, atomic structure, etc., but the beings there are not bound by them as we earthlings are. Think of it this way: Heaven contains all the same features as earth but also a whole invisible spirit realm, which allows the physical beings to fly, float, move instantaneously from place to place by merely thinking it, communicate telepathically, etc. Although skeptics will say that is a science fiction scenario, everything about God seems like science fiction—the whole creation story, the miracles mentioned in the Bible, the notion of eternal life, and the belief in a place of punishment/banishment called "Hell," being just a few examples. When one takes as an axiom, as believers tend to do, that all things are possible with God, then no scenario is too fantastic or sensational to consider as a plausible explanation for the otherwise unexplainable.

Whatever and wherever "Heaven" is, it is a place where humans cannot go by their own choice or efforts. Some are allowed to go there, or are taken there, by God or one of his other-worldly representatives (whom we will discuss shortly), but the human technology to get there does not and may never exist. Often, one of the first main objections that atheists and skeptics have to doubt the existence of God is the fact that we cannot show "where" God is. These doubters are locked into the mental prison of thinking that God is a one-dimensional being and thus must be "somewhere" and in only one place. The concept of omnipresence will not compute for them because they reduce God to having human limitations. Whichever of these theories or explanations is the case must square with God's third (figuratively speaking again) creation—the beings that He chose to surround himself with in Heaven. We commonly call them "angels," but they are in fact an innumerable assortment of beings, each different and special. These beings have been said to move from Heaven to earth and back from time to time. The Bible records several instances of these beings arriving on earth and presenting themselves to humans as God's messengers. Likewise, tales of visitors from the sky abound in pagan literature, oral traditions, and art. Scattered across the earth are many archaeological treasures that seem best explained by technologically advanced visitors from another world coming to earth and transferring the knowledge to humans. Even today, from time-to-time, people claim to be visited by these beings. Many scholars and casual observers alike have noted the similarities between biblical angels and what modern terminology holds to be "extraterrestrials" or "aliens." Most of the mysteries of history and archaeology can be explained by reconciling angels with extraterrestrials.

As we begin to explore this topic, we must be careful to remember that these heavenly beings are just as diverse as humans are. While they may share many similarities with each other, their characters and personalities are not robotic. They did not come off a cosmic assembly line. Each individual angel/alien is autonomous and can choose between good and evil, just as humans can. This is why there have been vastly different experiences recorded throughout history by humans who have seen or talked to these beings—some positive and some negative. We do not know what process God used to create them, whether it was similar to how He created Adam, whether He simply willed them into existence, or something else. We also do not know how many of them He created, whether He has ever annihilated any of them from existence, whether He created them all at once, or whether He created them uniquely and individually over a matter of "time." What we do not and cannot know about them is therefore much greater than what we do and can know.

The Bible refers briefly to a group of beings who lived on earth in ancient times called "Nephilim" (Genesis 6:4). The term translates to mean "giants." The biblical account of these beings says very little of any substance about them, but points out that they lived on earth at one time, or else visited earth regularly, and that they had sexual relations with human women who bore them children. For more information about them, we must turn mainly to pre-Christian literature, then extrapolate a uniform conception about them from these otherwise seemingly conflicting sources.

Ancient Greek mythology probably yields more clues on this topic than any other single written source. While it has commonly been taken for granted for at least the last thousand years that tales of the Titans, the lesser gods, and demigods are nothing more than fanciful, entertaining folklore, recent theories explaining ancient archaeological mysteries suggests that there may indeed be a historical basis behind the mythology. Hesiod's *Theogony* (circa. 700 B. C.), following earlier written works such as Homer's *Iliad* and *Odyssey*, as well as oral tradition, began the process of systematizing Greek mythology. The result was the story of super-human beings that created the universe and all that is in it. Containing too many characters (god-figures) to name much less discuss here, the gist of the work is this: beings not of this world came here from some other-worldly abode, were sent here by some other-worldly being greater than themselves, or were created here by one of the greater other-worldly beings—some deliberately, and some accidentally. They came to dominate earth and control the affairs of humans, partly because of their greater physical size and/or strength, and partly by their advanced technological knowledge. Unfortunately for humans, they were nowhere near perfect, as the Christian God is believed to be. Instead, they were lustful, moody, petty, vengeful, and capricious. They used and abused people as pawns in a cosmic game.

One problem with this Greek mythology is that it was constantly evolving from the time of Homer until the time of Christ and beyond. Homer's work did not pretend to be scripture; it was not meant to be taken that literally. It clearly taught great moral lessons—the main one being that hubris leads to bad outcomes—but it was not given to mankind on tablets of stone by the finger of God or any such thing. Hesiod's work was a more-nearly religious type, but was still malleable enough for later writers, such as the Roman poet Virgil, to build upon and alter. So many different versions of the Greek and Roman pantheon of gods had been developed over time, in fact, that more than a thousand years later, the Renaissance scholar Giovanni Boccaccio would still be cataloging them and trying to construct an encyclopedia about them. At first glance, this evolution of pagan gods in literature seems inferior to the near-rock solid scriptures of the Christian Bible. However, it would be wise to realize that while the Judeo-Christian scriptures themselves have not changed noticeably in nearly 2,000 years, interpretations of them have. So have decisions about which ancient texts deserve to be in the canon. Therefore, again, the Bible does not tell all there is to know on this important subject and can be filled out in a little more detail by supplementing it with the basics of Greek mythology. The basics are simply that a group of super-human beings at one time took up residence on the earth and interacted with humans, and while humans in ignorance naturally mistook them for "gods" because of their size, strength, and knowledge, they were in fact merely heavenly beings (angels/aliens) sent here from another world.

Greeks and Romans were not alone in developing a great mythological literature around these beings. The oldest civilization in the historical record, the Sumerians of Mesopotamia, had the same basic template. They wrote of the Annunaki, who can easily be considered the equivalents of the Nephilim or the Greek gods. Ancient Sanskrit writings from India also discuss gods with superhuman powers, as do old Germanic myths. Likewise, the Egyptians worshiped gods that seem to have been giants, because they built pyramids, obelisks, and statues on a scale that is out of proportion to all the surrounding landscape, while the technology to do so came from seemingly nowhere. Similar technologically advanced civilizations sprang up mysteriously all over the world—in the Middle East, Asia, Mesoamerica, South America, the Pacific Islands, and Europe—each with its own story of godlike beings who visited them in the past. When we inquire into the causes of the rise of civilizations in ancient history that built pyramids, ziggurats, towers, and other megalithic structures (such as the Moai on Easter Island, Stonehenge in England, Puma Punku in Bolivia, Baalbek in Lebanon, and the underwater cities of Yonaguni-Jima in Japan and Dwarka in India, just to name a few), no other explanation suffices. To say that humans went from an illiterate hunter-gatherer caveman lifestyle in one generation to suddenly a literate, civilized, and technologically advanced way of life in the next stretches the imagination beyond reach. The most rational explanation, therefore, is that humans had help from transplanted alien beings, some of whom were in fact already here on earth before God created the first homo sapiens.

So far, this explanation is not much different than those which any number of secular scholars and conspiracy theorists have posited for decades. Notably, Eric von Daniken's groundbreaking book *Chariots of the Gods* (1968) started a whole movement with this theme that has not subsided to this day, as evidenced by the recent popularity of television programs such as *Ancient Aliens* on the History channel cable network and any number of science fiction films. What this movement lacks, however, is any meaningful acceptance or assimilation of Christian theology into its inquiries or theories. That is not to say that they dismiss the Bible altogether, for they routinely delve into certain parts of the Old Testament and fit those passages into their theories. Ezekiel's wheel (see Ezekiel Chapter 1) gets a great deal of attention, because it seems to be a description of some kind of spaceship. Other passages, such as Enoch (see Genesis Chapter 5) and Elijah's mysterious removals from Earth (see II Kings Chapter 2), seem to imply the arrival and departure of spaceships, as does Jacob's vision of the ladder reaching into the heavens (see Genesis Chapter 28). Sodom and Gomorrah's destruction after the arrival of angels to warn the "good" people to leave (see Genesis Chapter 19) also implies extraterrestrial visitation, as do the several other cases scattered throughout the Bible where angels appear. But the spin that skeptics and scoffers generally put on these biblical tales is that the authors of the Bible simply mistook angels for extraterrestrials, superstitiously believing they were messengers from God. In their minds, there is no "God," for extraterrestrials, just like humans, are the product of evolution; they are merely further along in that regard than we are.

The truth is, however, that the beings mentioned in these Bible stories really were/are messengers from God, and God is not merely the most advanced life form on the evolutionary scale, but rather, as we have already established here, the creator who is above and beyond all the physical universe. Yet the full explanation is much more complex than this simple truth. In addition to the occasional appearances of "good" angels sent by God to earth on temporary missions bearing messages for humans, there are the countless other cases of "bad" angels who have been exiled here to earth and/or this region or dimension of the universe.

To understand how this came about, we must turn to the New Testament and specifically to the book of Revelation and square it with certain Old Testament verses and passages, such as those found in Isaiah, for example. In a vision, the author of the book of Revelation, John, sees war in Heaven. It is caused by the second most powerful being there, "Satan" (which means the accuser or deceiver, who was, according to the orthodox Christian interpretation, formerly known as Lucifer or the Light Bearer), who leads one third of the angels (his followers) in a revolt to overthrow God from the throne. Satan and his minions are defeated and cast out of the presence of God and down to earth, where they must live in exile until judgment day at some unspecified point in the future. Having but a short time (how long they did not know, nor do we), these former angels (now described as demons or devils in religious terminology, but known to secularists as aliens or extraterrestrials) must make the best of the bad situation they find themselves in.

One thing the Bible does not tell us is precisely *when* this war in Heaven took place (or could it even be yet to occur in the future?) The book of Revelation is most likely not a narrative of events arranged in chronological order. So the story of Satan's expulsion from Heaven may be set in the past or in the future. In order to get the rest of the Bible to make sense, however, the explanation has to be that it occurred in the past. It could be that God created a host of heavenly "assistants" to help carry out certain functions within the physical universe prior to the actual creation of that universe. If so, why? We have already established that God did not and could not "need" assistants. Therefore, it had to be because He had a plan in mind to use one group of living beings within his creation to make a statement or point to the rest of the living beings in his creation. That plan in its entirety is a mystery to us humans, of course, and must remain so until the time of the end when God chooses to reveal it. Meanwhile, we can know, or at least assume, snippets of it. Let us assume, for example, that Lucifer and many (if not all) of the other angels were privy to the scientific knowledge that went into the physical universe's creation. They had no power or ability to "create" new matter or life themselves, but they could manipulate the matter and life forms that God had created or was creating. Upon being banished to earth, they brought that knowledge with them and used it to corrupt the otherwise "perfect" creation of God. We will come back to this topic and have much to say about it later.

What state of existence these former angels had enjoyed in Heaven we do not know. Were they purely spirit-beings like we tend to visualize them because of common depictions that we have seen since childhood? Or were they physical beings like humans, but perhaps with extreme intelligence or at least vast capacity for it? Did they have physical bodies made of atoms and molecules that required air, food, and water? Or were they something heavenly that we can't comprehend because it has no equivalent here on earth, such as a unique blend of spiritual and physical? (It may be that in Heaven—a vastly superior place to earth—breathing oxygen may not be necessary and eating and drinking may not be necessary, for the inhabitants may enjoy alternative ways of absorbing, ingesting, or acquiring the essentials for life.) Or maybe they had the ability to mutate to fit whatever environment they found themselves in? We do know that they were not all exactly the same. They had their specific roles in Heaven, and they were arranged in a hierarchy, with Lucifer being at the top. They probably came in a great variety of sizes, shapes, colors, and textures. Whatever their characteristics and appearance in Heaven, some apparently have been able to exist as disembodied spirits on earth. Otherwise, there is no good scientific explanation for "Legion" (see Mark Chapter 5) and similar cases of demonic possession mentioned in the Bible.

Upon being expelled from Heaven, the former angels were reduced to existing within bodies which had compatibility with the atmosphere and climate of earth. Earth was clearly not ideal for them, for it limited their ability to use whatever gifts and intelligence they had, which were designed with heavenly, not earthly, functions in mind. This downscaling did not eliminate those gifts altogether but rather made them subject to all of the laws of the physical universe, which they may not have been subject to in Heaven. On earth, they were/are probably, for example, able to communicate with each other telepathically, and likely able to manipulate physical objects through mind control (telekinesis). But it is unlikely that they could/can perform time travel or engage in teleportation, for these are the gifts reserved for the non-bodily beings of the heavenly realm. (Even the good angels who still inhabit Heaven cannot defy the laws of space and time once they leave Heaven and enter the physical universe.) Thus, they had/have to move through physical space and time just as humans do. They may well have the ability to move much faster than we do, thanks to their advanced technology, but they cannot surpass the speed of light. Likewise, they are not immortal, although their life span may far exceed that of humans—so much so, in fact, as to appear to us to be practically immortal.

This raises a question: when God expelled Lucifer and his followers from Heaven, how did they get to earth? Did they come in spacecraft built in Heaven but designed to navigate the physical universe, which God let them fly away in, as it were? Or did God instruct the good angels to transport them in such vehicles to earth, drop them off, and leave them with no form of transportation to escape? If the latter, then these fallen angels subsequently built their own interplanetary transportation devices from materials found here on earth. This is not beyond the realm of possibility. If it happened this way, we have no way of knowing how long they were in the building process, for it would have entailed discovering and gathering building materials. It is possible that they were still engaged in it when the first humans were put on earth. This could potentially explain the Sumerian story of the Annunaki using humans as slaves for mining gold and other raw materials for them.

Whatever the case, God, in his extraordinary mercy, apparently allowed these former angels a measure of freedom to move about as they pleased, except they could not re-enter the abode of God himself from whence they had been cast out. But He did allow them to navigate through the physical universe to some extent, even if only the parts in the immediate vicinity of earth. Finding no life anywhere else, however, they have occupied earth or stayed close to it ever since, except some of them may be on a never-ending exploratory mission to find another planet capable of sustaining life. If so, that could mean they were not privy to that information in Heaven about whether God created life on some other planet, and they want to find out. Or it could mean they were indeed given that information, but they either have not found that planet yet or have no ability to get there with the type of spacecraft they are using. We tend to think of these extraterrestrial beings as having an unlimited ability to travel, but the truth is, they are indeed limited by the vastness of space. In short, they were consigned to an existence like that of humans, only they had a big head start, in terms of the knowledge they already possessed, in making the best of it. While here, they imparted some knowledge to humans, but always did so for their own selfish purposes, never out of altruistic motives. They did not always work together and agree on their strategy, tactics, or objectives, either, so they frequently engaged in internecine power struggles with one another, and they were happy to use humans, animals, and anything else as pawns and weapons against one another. This made them appear to be capricious and fickle, which in turn instilled fear of them in humans. They may have lived here or in their spaceships for eons before God created other biological life to surround them on earth. Or they may have lived here only since the creation of other life on earth.

One likely scenario for how this story unfolded is this: God gave Lucifer and his angelic comrades the job of helping set up the conditions on earth to sustain life. Once God created life, He then gave them the job of scientifically modifying those life forms. In the midst of these scientific "experiments," Lucifer came to discover two things: one, that the power to create "new" life forms (species) from existing ones was extremely ego-aggrandizing; and two, that he could make "bad" species as well as good ones. At first, he had no reason or inclination to produce bad ones. In time, however, as he saw God's plan begin to take shape, he found a reason.

Rather than go any further with this story here, let us save it for the next few chapters where we will plug it into the chronological narrative of Genesis.

1

THE EXPLANATION
GENESIS Chapter I

A DAY IS AS A THOUSAND YEARS

Although God (the Elohim) is a time-less being and is not subject to the restrictions that time places on all physical things in the universe, He is apparently an avid time-keeper. He has no need of keeping time for his own benefit, of course, because He is eternal and infinite, having no beginning or end. That's why "a day is as a thousand years, and a thousand years as a day" (II Peter 3:8) to him. The emphasis here must be placed on "as." It means a "day" might as well be a thousand years—or a million years, for that matter—because from God's perspective, time is relative; it is merely a construct that He created which allows physical beings to measure finite pieces of eternity, and it is only necessary when used within a physical, spatial universe. To create an orderly physical universe for the benefit of the beings that would inhabit it, therefore, God created time. To put the concept of a day being the same as a thousand years to God in proper context, we will deliberately substitute the word "millennium" for the word "day" in the subheadings in this chapter. It is vitally important to be clear that this word is not to be taken as a literal measurement of time, however, but rather as a way to emphasize the immeasurability of the time that passed during each "day" of creation. And that brings us to discussing the day-by-day (or millennium-by-millennium) creation story of Genesis Chapter 1.

> 1 In the beginning God created the heavens and the earth. 2 The earth was without form, and void; and darkness was on the face of the deep. And the Spirit of God was hovering over the face of the waters. 3 Then God said, "Let there be light"; and there was light. 4 And God saw the light, that it was good; and God divided the light from the darkness. 5 God called the light Day, and the darkness He called Night. So the evening and the morning were the first day.
>
> 6 Then God said, "Let there be a firmament in the midst of the waters, and let it divide the waters from the waters." 7 Thus God made the firmament, and divided the waters which were under the firmament from the waters which were above the firmament; and it was so. 8 And God called the firmament Heaven. So the evening and the morning were the second day.
>
> 9 Then God said, "Let the waters under the heavens be gathered together into one place, and let the dry land appear"; and it was so. 10 And God called the dry land Earth, and the gathering together of the waters He called Seas. And God saw that it was good. 11 Then God said, "Let the earth bring forth grass, the herb that yields seed, and the fruit tree that yields fruit according to its kind, whose seed is in itself, on the earth"; and it was so. 12 And the earth brought forth grass, the herb that yields seed according to its kind, and the tree that

yields fruit, whose seed is in itself according to its kind. And God saw that it was good. 13 So the evening and the morning were the third day.

14 Then God said, "Let there be lights in the firmament of the heavens to divide the day from the night; and let them be for signs and seasons, and for days and years; 15 and let them be for lights in the firmament of the heavens to give light on the earth"; and it was so. 16 Then God made two great lights: the greater light to rule the day, and the lesser light to rule the night. He made the stars also. 17 God set them in the firmament of the heavens to give light on the earth, 18 and to rule over the day and over the night, and to divide the light from the darkness. And God saw that it was good. 19 So the evening and the morning were the fourth day.

20 Then God said, "Let the waters abound with an abundance of living creatures, and let birds fly above the earth across the face of the firmament of the heavens." 21 So God created great sea creatures and every living thing that moves, with which the waters abounded, according to their kind, and every winged bird according to its kind. And God saw that it was good. 22 And God blessed them, saying, "Be fruitful and multiply, and fill the waters in the seas, and let birds multiply on the earth." 23 So the evening and the morning were the fifth day.

24 Then God said, "Let the earth bring forth the living creature according to its kind: cattle and creeping thing and beast of the earth, each according to its kind"; and it was so. 25 And God made the beast of the earth according to its kind, cattle according to its kind, and everything that creeps on the earth according to its kind. And God saw that it was good. 26 Then God said, "Let Us make man in Our image, according to Our likeness; let them have dominion over the fish of the sea, over the birds of the air, and over the cattle, over all the earth and over every creeping thing that creeps on the earth." 27 So God created man in His own image; in the image of God He created him; male and female He created them. 28 Then God blessed them, and God said to them, "Be fruitful and multiply; fill the earth and subdue it; have dominion over the fish of the sea, over the birds of the air, and over every living thing that moves on the earth." 29 And God said, "See, I have given you every herb that yields seed which is on the face of all the earth, and every tree whose fruit yields seed; to you it shall be for food. 30 Also, to every beast of the earth, to every bird of the air, and to everything that creeps on the earth, in which there is life, I have given every green herb for food"; and it was so. 31 Then God saw everything that He had made, and indeed it was very good. So the evening and the morning were the sixth day. [NKJV]

Taken literally, the story of the creation of the "heavens and the earth" and all that is in them lasts precisely six days. During these six days, God did his work of creating all that was created. On the seventh day, He rested. The Bible does not say what God did on the eighth, ninth, or tenth days, or any other day, for that matter, after the seventh. We therefore cannot know exactly what, or what all, God has been doing for all these centuries (or eons) since the creation, nor can we know what He is doing right now. The Bible gives us tiny slivers of information about certain things God has done at certain moments since the creation, but there is far more that it does not tell us than what it does tell us. We will return to this topic shortly. Whether the seven-day creation story of Genesis is meant to be taken literally as seven 24 hour days (where each day equals one complete rotation of the earth) is an important question. Although the simplest way to interpret the story is to take it literally, we must inquire as to whether it is the most rational, and thus the most prudent, interpretation. One problem with the literal interpretation is that the narrator of the story (ostensibly Moses) starts counting the days before the sun, moon, or stars are created. How could it be determined how long a "day" was if there was no sun or other lights in the sky by which to keep time? Was God keeping time in Heaven? Had He predetermined that a day would be 24 hours even before He created the sun and set the earth in motion spinning on its axis in orbit around it? Actually, maybe so. Before we automatically dismiss that possibility as absurd, let us consider it. God would have known that only a certain speed of rotation would allow for harboring life on a planet this size. This particular speed helps keep the earth stable; it gives the earth its perfectly-tuned wobble that produces the seasons and keeps more than half of the planet in livable temperatures for more than half of the year. It also gives earth just the right amount of gravity to keep things from either being crushed or flying off into space. And to some extent it may even help earth maintain a certain regular, elliptical orbit rather than allowing it to veer too far from the sun or get too close to it. So it is possible that God started keeping time immediately upon the Big Bang, even before our Solar System was set in place in its current form. Besides, if we suppose that the earth was modeled after "Heaven," the abode of God, this hypothesis can make sense. Let us explore this possibility by juxtaposing it against other common beliefs and ideas about primordial origins.

MILLENNIUM ONE

Creationists typically read Genesis Chapter 1 chronologically and get bogged down in a quagmire of non sequitur events. Consequently, we cannot agree with one another on various points. So called "Old Earth" Creationists argue that the earth is billions of years old, just as secular scientists believe, based on the fact that what happened in Genesis 1:1 gives us no clue how much time elapsed before Genesis 1:2 occurred. Nor does the first half of Genesis 1:2, they say, tell us how much time elapsed before the second half of Genesis 1:2 occurred. "Young Earth" Creationists tend to be Fundamentalists (meaning they take everything in the Bible literally), but even if they are not Fundamentalists on all points in the Bible, they definitely believe that creation happened in six literal, successive 24-hour days. They disregard all evidence to the contrary and offer conflicting pieces of scientific data to argue their point, although their opinion rests mainly on faith in God's omnipotence rather than on science. (They claim there is some evidence, for example, of a great global flood a few thousand years ago, just as told in the Bible.)

The Old Earth Creationists' point about the length of time between Genesis 1:1 and Genesis 1:2, and the first half and the second half of verse 1:2, requires examination here. Reading chronologically, they naturally put the creation of the literal "earth" in verse 1 prior to the Big Bang in verse 3. Consequently, they talk of how much time elapsed during which the literal "earth was without form, and void; and darkness was upon the face of the deep" in the first half of verse 2 prior to "the spirit of God was hovering over the face of the waters (oceans)" in the second half of that verse. [Rather than using the past tense verb of the old KJV that implies a one-time event, "moved upon," the NKJV uses a verb that shows continuing action—"was hovering over."] Then they cannot adequately explain how God created light in verse 3 after creating the earth, except to say that perhaps the newly formed planet was covered in such dense gases, smoke, clouds, and/or whatever else as to block out all light from the sun.

This method of interpretation, which is not necessarily and demonstrably wrong, could be arrived at in two different ways. One is by reading the creation story this way: "In the beginning, God created both the heavens and the earth, but let's ignore the creation of 'the heavens' for the moment and focus just on the creation of 'the earth'..." At that point, the narrator proceeds to look at the rest of the creation story from the vantage point of earth, as if from inside its atmosphere looking upward and outward. That interpretation allows for light *seemingly* to be created *after* the heavens and the earth were created, for it would have appeared that way from an earthly vantage point. It also allows for the most common sense reading of verse 14 later in the story, to which we will return shortly. The second way to arrive at this method of interpretation is by distinguishing between "the heavens" of verse 1 with "Heaven" in verse 8, with the former referring to the sky or atmosphere of the earth and the latter to all space beyond the earth's atmosphere. When reading Genesis 1:1 this way, we get the narrator referring to the Big Bang not as if he were looking at it from outside or beyond it as God would have seen it but only from the vantage point of an observer on earth. Although "the heavens" as used in verse 9 would be consistent with its usage in verse 1 if interpreted with this method, its usage later in verses 14, 15, and 17 would not be. In those later verses, "the heavens" instead seems to be used synonymously with verse 8's "Heaven," making for a quite muddy mix that probably renders it incorrect. Verse 9, however, gives us trouble if we *don't* interpret it as the sky or atmosphere, so either way we get a muddy mix.

Even though for all its difficulties the first Old Earth Creationist model above may be plausible, whether it is the correct explanation is debatable. Modern science offers an explanation that is, of course, quite different. It holds that the Big Bang was an explosion of matter that produced light, heat, and shock waves from which the earth, the moon, the sun were created, along with trillions of other stars, planets, asteroids, comets, and space particles scattered through billions of galaxies spread in all directions through space instantaneously. Moreover, it implies that this creation-by-explosion did not stop, but it continued for eons (billions of years as measured in earth time), and in fact is still happening because the universe is still expanding! The creation, therefore, was not a single event lasting 6 literal 24-hour days at some point in the past, but is rather an ongoing process.

The theological problems with this scientific theory are obvious. First, this theory conflicts with the apparent past tense used in the Genesis story, which conveys the notion that the creation was a one-time, historical event that had a definite ending point—6 "days" after it began. And second, it reverses the order of the story told in Genesis, making verse 3 come before verses 1 and 2. But upon careful examination, these problems which seem to reveal conflict between science and theology can be made to disappear quite easily. We merely have to broaden our ability and willingness to grasp the big picture of Genesis, rather than get hung up on a verse-by-verse, sentence-by-sentence, and word-by-word reading of it. Although this is especially true when considering English translations of Genesis, it is also true even if we were to try to go line-by-line in the original Hebrew.

To square this modern scientific explanation with the Genesis story, we simply must assume that the Big Bang occurred not in verses 1 or 2, but rather in verse 3, at the moment God said, "Let there be light." To arrive at this conclusion requires only one of three things: A) that we move the order of a couple of verses around, B) that we change the translation of a particular phrase within the first two verses of Genesis to match what seems to be the observable reality, or C) we accept the possibility that Bible verses can and do sometimes have two distinct but equally true interpretations at once. To do any of these things may seem like sacrilege to Fundamentalists, but it still requires faith to believe that an outside-of-creation force (God) was responsible for the Big Bang and all that we observe in the universe. And, more importantly, it keeps all things created on the first "day" of creation together as a unit, all happening on the same day, regardless of the order in which they occurred on that day. It does not, therefore, require that we move second or third day events ahead of first day events, which would absolutely bring the whole framework of the story crashing down. Although either of these two options, A or B, would be an improvement over a Fundamentalist interpretation, it is possible that applying both of them will actually produce a more rational explanation of what happened at the first moment of creation. Then adding option C to this mix will give the most rational explanation of all.

Concerning point B, if we translate the phrase "the heavens and the earth" to mean "all space and matter" (i.e., all properties of the physical universe, including time and motion), we find a reasonable way to get the rest of Genesis chapter 1 to make sense. To explain, "the heavens" really refers to all of the vastness of the universe (space), and "the earth" really means all of the physical matter within that vast space. So, to translate verse 1 this way, we get: "In the beginning God created all space and matter." This method of translation creates a problem in verse 2, unless that verse is likewise reworded for clarity. We must understand, "The earth was without form, and void . . .," to mean that all matter was void of any set form (it was malleable in the hands of the creator, in other words). Continuing verse 2 with this same method of translation, we must interpret the clause "and darkness was on the face of the deep" to mean that all of this matter existed within the darkness and emptiness of space. The next sentence in verse 2 is the most problematic of all. When it says, "And the Spirit of God was hovering over the face of the waters," it seems to be referring to the H_2O that makes up the majority of the surface of the earth. At least that is the common sense way to read the sentence, and we must admit that any time there is a common sense way to read a biblical passage, it might well be correct. But in this case, is that really what it means? Not if we interpret it consistently with how we interpreted the sentences and clauses before it. If we translate "waters" as a fluid or malleable synonym for "earth" or "matter," it makes more sense: "The Spirit of God (the creator) held this malleable matter in his hands" to shape it, form it, and turn it into something solid. He was holding it, in other words, and brooding over what exactly He was about to do with it. (That does not mean He was "wondering" what to do with it, because that would make God appear indecisive, and clearly God cannot be indecisive or else He is not really God. So to say He was brooding over what He was about to do implies that He was working within the time constraints of the physical universe He had just created in order to make events unfold at precisely the right time, speed, distance, force, etc.)

Applying option C to verse 2 yields the possibility that the narrator was telling us two things at once. First, as mentioned, he was informing us of what happened before the Big Bang as God held all matter and space in his hands, but he was simultaneously forecasting the actual chain of events that was about to unfold on the earth once created. We will return to this chain of events on Day 2 of the creation story. But for now, let us stay focused on the first moment of creation.

Modern science postulates that the universe and all that it contains began from an infinitely dense but sub-atomically small point, which is precisely what this explanation in option B suggests. Modern science cannot tell us what brought that point of infinite density into existence, how long it existed prior to the Big Bang, or what caused it to explode when it did. Some scientist-philosophers (called cosmologists, cosmogonists, theosophists, and perhaps some other things) speculate about first causes (ontology), but they cannot prove or disprove a speculation, so the Genesis story is just as possible and likely as any non-theological explanation that any secular scientist can offer—such as that there are many universes with many creators, the universe is its own creator, etc. The explanation, therefore, is that God held this malleable substance of infinite density in his hands and shaped it into the universe we know today through the process of the Big Bang and its natural physical consequences. This explanation makes much more sense than reading verses 1-3 straight through chronologically, because it eliminates confusing contradictions.

Concerning the natural physical consequences, the best probable explanation is that in the immediate moment ("day"?) after the Big Bang, all the matter in the universe

—galaxies, solar systems, individual stars, planets, asteroids, comets, etc.—was flung in all directions from a central point. The earth was formed at this moment as just one of trillions upon trillions of chunks of matter. It was not simply created intact in its current form, or even remotely like it, all at once, as Young Earth Creationists contend. It lacked a settled, stable form; it lacked life, and more importantly, the prerequisite light for giving rise to life. The earth underwent a cooling and settling process that lasted millions of years, which might very well be Day 2 in the creation story—and we will have more to say about this shortly. During this time, the planet's inner and outer temperatures cooled gradually, the oceans stopped pitching and boiling, toxic gases dissipated in the atmosphere, a nitrogen-oxygen-carbon dioxide mix came to dominate, and land appeared. During the same time, the earth settled into a regular rotation that would last 24 hours, into an orbit around the sun that would take 365 days, and into an axis-wobble similar to what it has today that causes change of seasons. The moon simultaneously settled into a regular orbit, steadying and regulating the earth in the ways mentioned above.

How long this process or "day" took, secular scientists and Old Earth Creationists can only speculate. Billions of years is a nice, round, ballpark, educated guess. However long it took, ostensibly, throughout the process the "Spirit of God was hovering over the face of the waters," as the NKJV puts it. But in order to remain consistent with the re-interpretation offered thus far in this exegesis, we must presume the clause to mean that the Spirit of God *continued to hover* over his ongoing creation. This method of translation informs us that God did not rest after the Big Bang but rather remained actively involved in the creation process for however long it took. According to this interpretation, whether it was 6 literal earth days or 4 billion years or more is a moot point. The real point is that God "*was* hovering over," meaning He was *continually* observing the creation process He had started. The emphasis here is on the past-tense verb, because even trying to be open-minded to science, Old Earth Creationists still read the Genesis story as a historical rather than an ongoing event. The question naysayers of this explanation will undoubtedly and rightly raise pertains to day 7 when God rested from his labors. Indeed, at first glance this explanation does not seem to jibe with the Genesis account on that point. We will address that point when we get to day 7, but for now let us continue focusing on Day 1. To recap Day 1, Genesis 1:1-3 should read: "In the Beginning, God caused the Big Bang which created all space and matter, and the matter was and is being spread throughout the space, forming the universe as we now observe it."

Moving into verses 4 and 5, the narrator tells us that God saw what He had created on that first "day"—light, space, and matter—and pronounced that it was good. Ostensibly, this means that while the creation process that He set in motion continued, the first phase of it came to an end by divine declaration. Thereupon, He distinguished between the darkness and the light, giving each a name, "evening" and "morning." This, of course, raises another theological dilemma, because it seems to contradict the Old Earth Creationist and secular scientist interpretation already mentioned. So how can we square the evening plus morning equals a day terminology with observable reality? Let's look at our options: it may be that A) God set the earth in a regular 24 hour rotation around the sun immediately upon their creation in the Big Bang; B) the abode where God was/is while performing the physical creation also had a regular pattern of going from night to day, night to day, night to day, etc., in which case verse 5 does not refer to the rotation of the earth but rather to the pattern of "Heaven" which would ultimately be applied to earth as well; C) evening and morning are really just allegorical terms for two phases of the Day 1 creation process; or D) a combination of options B and C provide the best explanation.

To evaluate the merit of each: option A is the Young Earth Creationists literal interpretation, which does not allow room for the scientific theory of the long, slow cooling process already mentioned. Option B is a feasible possibility based on the explanation put forth earlier of who God is and where He resides. If we assume the narrator is telling us what God observed from Heaven, using the heavenly timetable as the template from which the earth's timetable would be copied, then this option makes sense. It really makes perfect sense, however, when combined with Option C. If we consider that the earth was created as a giant chunk of matter flying through space for millions of years before beginning to settle down into a regular rotation and orbit within the solar system, we can easily visualize two distinct phases of the Day 1 creation process. Phase I was the time during which the earth's atmosphere, land, and water had not yet separated into their current forms. The whole earth was just one huge, boiling cauldron with no distinction between land, water, and sky. This could easily be likened to "evening" (a time of darkness), and it would explain why evening must precede morning in the story the narrator is telling: it was the time when the atmosphere began to take shape and allow light to penetrate to the surface even as the surface itself was likewise taking shape. It is important that we provide some such explanation for this otherwise curiously backward-looking order in which each "day" appears in the story, and this is probably the best one. It squares with science without destroying the integrity of the big picture of the Genesis narrative. We will see shortly how this same interpretation will allow for a common sense rendering of the events of days 2 through 6 as well.

MILLENNIUM TWO

On Day 2 of creation, God separates "Heaven" from whatever "waters" were/are beneath it. Verses 6 through 8 here seem to refer back to the second sentence of verse 2, in which, according to option C, God hovered over the earth specifically while continuing the broader universe's creation. Skipping verses 3-5 (the Big Bang verses), verses 6-8 seem to connect with verse 2, because both passages speak of "waters." Just as we interpreted verse 2 as having a duel meaning, verses 6-8 could have that as well. If we take "Heaven" to mean all of the space that separates matter from matter in the universe, we see that the distances are growing, that galaxies and solar systems and planets are settling into rotations and orbits, and that the earth is part of that process that began immediately after the Big Bang in verses 3-5. If we take "Heaven" to mean the sky as it appears from the surface of the earth, however, we see in these verses that the long slow cooling process had entered stage two, in which the earth began dividing and settling into two basic parts—oceans and atmosphere.

The term "firmament" that is used five times here is troublesome. Translated from Latin, which was itself taken from a Hebrew term meaning something like bowl-shaped, dome, or sphere, this English word implies something "firm," solid, and/or unmoving. Yet the context in which it is used in the creation story implies something that simply separates some parts from other parts or divides individualized parts from the whole. When applied to the universe as a whole, it could be interpreted as whatever force or forces hold chunks of matter together as planets, solar systems, or galaxies, while simultaneously pushing the various chunks away from each other (gravity, strong nuclear force, weak nuclear force, electromagnetic force, etc.) In other words, even after the Big Bang, there had to be a "firmament" or set of physical laws put in place to make an orderly universe that is constantly expanding yet which prevents the chunks from coming apart. When applied to the earth itself, the "firmament" could be construed to mean that which separates the surface area from the atmosphere above it. One definition of the term actually implies that the denser surface holds up the gaseous sky above it while at the same time holding it within its gravity so that it cannot escape and dissipate into space. Either way, verse 8 says that "God called the firmament Heaven" (with a capital H), which seems to be the same as "the heavens" in all other verses in Genesis Chapter 1—except for verse 9—provided we do not subscribe to the Old Earth Creationists' second method of interpretation already mentioned. The separation of "dry land" from the "waters under the heavens" in verse 9 implies the separation of land, oceans, and atmosphere, not of earth as a whole from "the heavens" as used elsewhere in Genesis.

Whatever this firmament is, it is vitally important to the creation story, because it is the sole thing to be "created" on the second day. The only way to make this mysterious piece of the puzzle fit into the big picture is to assume, as mentioned above, that it is the whole set of physical laws that were put into place after the initial Big Bang. In other words, it is the thing or things that hold(s) the parts together atomically as masses while pushing all these various masses away from each other over the vastness of otherwise empty space at the same time. That would be an important enough thing to deserve its own "day" in the story! And it can be applied not just at the macro level of the universe but also at the micro level of the earth. It was the process of the earth simply taking shape and settling into a regular orbit, rotation, etc., as previously mentioned.

MILLENNIUM THREE

Verse 9 introduces Day 3 of the creation story. Here God separates the land masses from the oceans and various other bodies of water. This is merely the continuation of the settling process that occurred all through Day 2. It is the last part of it, in fact, but ironically the first place in the story where God voices an opinion, or renders a judgment, about his own work after observing it. He declares his satisfaction with his creation, saying it is "good." The Hebrew term from which we get the word "good" in our English translation implies something more than that; it implies that God looked at his work and said it is "right," it is "as it should be," and it is "just as I intended and exactly what I wanted." That declaration is the first time we see God personified in the creation story. Prior to that, God was purely the omnipotent, supernatural, and metaphysical Elohim. Here He seems to have a human trait. He decides to express himself verbally. To whom was He expressing himself, though? To heavenly beings that already existed beyond and apart from this physical creation? Why was He expressing himself? To give those heavenly beings the opportunity to share in his satisfaction?

Or perhaps because He knew that someday humans would record his first utterance and somehow benefit from it? Of course, there is no way to know the mind or motivations of God for sure. What we can know, however, is what has already been explained—God (the Elohim) cannot be in need. He did not "need" to express himself anymore than He "needed" to create the universe. But if God's physical manifestation, the Lord (Yahweh/Jesus), was the one who voiced approval for his work to the other celestial beings and to the Elohim (the Father), it makes sense.

There seems to be a great irony in God expressing himself for the first time at this particular point in the creation story, because what comes next in the story seems much more important. It seems important enough, and in fact *different* enough, from the preceding events to warrant a bold proclamation of satisfaction. It is the creation of life. Oddly, the creation of life is mentioned almost casually in verse 11, as if it is a natural consequence of all that has come before it, as if it were bound to happen, and thus should not be considered any more special than the separation of the land from the seas. To make sense of this curious timing, however, requires a minimal effort. We simply must take God's expression of satisfaction not as his patting himself on the back and saying childishly, "Look what I just did!" but rather as an update of the progress made so far in his creation plan, of which the introduction of life would be the most important part. In that sense, the expression was really just an announcement of the creation of life about to come. He was saying, in essence, "All I have created so far is good for harboring physical life, which will be my greatest creation of all."

The introduction of life in the creation story begins, not surprisingly, with vegetation rather than with animals. This is in perfect harmony with what science postulates (if we disregard microbial life forms, or if we do not try to make too fine a distinction between different microbes). Although there are whole subfields of science devoted to the study of microscopic forms of life, we need not get bogged down in that topic at this time. Keeping the discussion therefore in the realm visible to the naked eye, we can safely say that before there could be moving parts to the puzzle (animals), there had to be stationary parts (plants). The narrator of Genesis calls these plants "grass," "herb(s)," and "fruit trees," each bearing the "seed" of its own reproduction/multiplication. Like the cosmological, geological, and chemical processes mentioned in Day 1 and Day 2, this sprouting of life must have taken millions and millions of years. One evidence of this fact is the curious wording of Genesis: "Let the earth bring forth . . .," rather than "Let there be . . ." Although it is certainly possible, as Young Earth Creationists would have it, that God could have simply spoken thousands of varieties of grasses, vegetables, and fruits into existence with a word, it defies human ability to understand or explain how such a command might have produced such a large and diverse body of life in a single 24-hour day. It seems more plausible that, having set a chain of events in motion from the moment of the Big Bang which led to the precise conditions necessary for the emergence of life, life in fact emerged right on schedule through natural processes.

Fundamentalists will of course protest on the grounds that, if that were the case, scientists should be able to replicate these chemical processes in a laboratory, but they have not been able to as yet. They will also protest that this theory reduces God to the Deistic clockmaker. Concerning their first objection, rational thought requires us to acknowledge that just because scientists haven't figured out how to do it yet doesn't mean they won't someday figure it out. On their second objection, it is not necessarily true that the theory reduces God's role in the creation. A plain reading of the Genesis text shows God being "there" during every step of the creation process, overseeing it, expressing approval of it, and in fact being at least partly hands-on about it. There is no reason to discount his presence at the moment of the creation of life, therefore. To reconcile the theology with science merely requires that we acknowledge that, whatever the missing "ingredient" necessary for life was, God supplied it at the moment when the earth by course of nature was finally equipped to make use of it. Once supplied, however, the long, slow process of evolution kicked in. God did not physically have to change the molecular structure of grass into weeds, weeds into briars, briars into flowers, etc. He built into the fabric of all flora the ability to adapt to the environment and reproduce with successive modifications over time.

The very mention of the word "evolution" above will likely disgust many Young Earth Creationists and/or Fundamentalists. Before dismissing the argument, however, they should consider the following. While it might have happened the way just described, a more likely explanation is that God actually created three distinct "kind(s)" of plants on Day 3, just as the Bible says—the grassy kind, the vegetable kind, and the fruit kind—and from each one descended an incalculable number of varieties through a long evolutionary process. "Grass" can be construed as a generic term for all plants that are typically considered inedible from a human perspective. Moss, algae, and many different flowering plants, groundcovers, shrubs, and trees might be in this category. "Herb(s)" probably refer to all vegetables that humans typically eat, and the meaning of "fruit tree(s)" is obvious. A few problems with this theory are that 1) some species of plants seem to fit into none of these categories neatly but might fit into two of them simultaneously, 2) this three-tier categorization may or may not account for fungi and other hard-to-classify life forms (i.e., is it a plant, an animal, or something else entirely?), and 3) there seems to be a hierarchy here in which the trees appear more "evolved" than the shrubs and the mosses because they are larger, denser, live longer, etc.

What then is the advantage of believing that God created 3 "kinds" of plants rather than just starting one kind of plant life and letting it evolve into every kind we have today? Two things: 1) it provides a rationale for why the narrator listed three distinct kinds of plants rather than just simply proclaiming matter-of-factly that God said, "Let the earth bring forth vegetation"—which helps defuse the secular scientific belief that life just spontaneously emerged with no God in the picture and then proceeded to evolve through natural selection rather than divine orchestration, and 2) it is more consistent with what science tells about the emergence of life than is the Young Earth Creationist belief that God created thousands of separate varieties one at a time through a hands-on process in a 24-hour period. To answer in advance a criticism that some will undoubtedly have of this thesis, it should be noted that just because God *didn't* carry out the creation with a hands-on process for each of the many thousands of varieties doesn't mean He *couldn't*. Certainly, He could have. If that is the case, then why not just assume that is what He did? Because, again, the scientific evidence suggests otherwise. This point will be reiterated when we get to verse 20 and consider the creation of animals.

MILLENNIUM FOUR

Meanwhile, verse 14 introduces Day 4 in the creation story. Of all the verses covered thus far, this one potentially could be the most problematic, because it seems to get the chronology wrong. Clearly, the creation of the stars, galaxies, solar systems, and planets did not happen *after* the creation of plant life on earth. To believe otherwise defies not only our scientific knowledge but plain common sense. Although an omnipotent God could have sustained life on earth by some other means while waiting to put the sun in place overhead, to assume that is what happened is not prudent. These Day 4 events instead seem to be a continuation of what happened on Days 1 and 2—the long, slow, settling process. If so, why would the narrator have separated those events rather than lumping them together? Because he wanted to place an emphasis on the perspective from which the events are being viewed. As already mentioned, the events of Days 1 and 2 make God look like an impersonal, metaphysical being who was performing a creation from the outside and looking down on it. Logically speaking, that must have indeed been the case, and it helps support the idea of an all-encompassing Elohim preceding the creation. On Days 3 and 4, however, the focus shifts to earth. Here God is performing the remainder of the creation from the inside, as if standing on the earth's surface and looking outward at it. That, too, could be logical, assuming the idea that one manifestation of God (the physical Lord/Yahweh/Jesus) was in charge on the inside while the spirit Elohim (Father) remained outside the creation.

Indeed, reading verses 9 through 19 this way allows the story to continue flowing chronologically; simple plant life emerged while the settling of land, water, and atmosphere was still underway. From the vantage point of Yahweh or some other celestial being standing on the earth observing or participating in the continuing creation, the sky had not sufficiently cleared prior to Day 4 to allow the sun, moon, or stars to be visible. Those bodies of light were there, and they gave off light which illuminated the surface of the earth well enough to allow plant life to emerge, but the sky was so overcast with steam, vapor, fog, clouds, etc., that the individual sources of the light could not be identified. On Day 4, the great clearing occurred. Blue sky and the sun became visible in the day, and the moon and the stars became visible at night. So by the end of Day 4, the earth had settled into the regular pattern that we all recognize and take for granted today. As it did, the evolution of plant life from Day 3 continued and accelerated.

At first glance, these events seem anticlimactic, as if they are not significant enough to deserve their own "day" of creation. Yet, without this extra day inserted after the creation of plant life in verse 11, enough time would not have elapsed to allow evolution to set the stage for animal life to emerge in verse 20. Although God himself didn't "need" that time, everything within the creation apparently did because of the firmament principle already mentioned. So Day 4 allows us to see the slow passage of time without much else happening to distract attention from it. Although time passes just the same over the first three days of creation, there seems to be enough "action" or "activity" during that time to prevent us from really grasping just how slowly this process moved. Although a day and a thousand years are the same to the eternal Elohim, humans as a whole have very little patience and thus feel an inherent need to count and account for the passage of time. That is why the events of Day 4 seem insignificant—we want concrete action, not the dreadfully slow process of evolution!

MILLENNIUM FIVE

Thankfully, Day 5 satisfies our need for action. In fact, other than the Big Bang of Day 1, there is more activity to be found on Day 5 than any other day of the creation. Here, in verse 20, God creates animal life. Finally, the first conscious, mobile, living organisms show up on earth. Two facts are interesting and important about the narrator's presentation here: 1) as in the case with the introduction of plant life, he says again, "Let the earth bring forth . . .," which seems to agree with the evolutionary theory, and 2) that animals appear first in the "waters" and "above the earth." Indeed, swimming and flying creatures precede the terrestrial animals, which can be squared easily with what secular biologists tell us about the origins of life. They assert that life began in water, although their details of the evolutionary progression of species differ from the biblical account, and thus from this explanation in places. The standard secular scientific theory holds that a single cell organism spontaneously came to life in water several billion years ago, and from it evolved all other animal life forms. Although it is conceivable that life arose this away, it is *just barely* conceivable. It frankly stretches the imagination to believe it. Essentially, the theory skips over "how" life spontaneously arose and jumps directly into the evolution of that life afterward. Unless science can discover some natural method by which non-living things come to life, the theory is no more than science fiction. Until such time, it is more plausible to believe that a spirit-being which exists apart from the physical creation—i.e., God (the Elohim)—was/is the originator of physical life.

Scientists have, of course, tried for decades to replicate in a laboratory the moment of life's inception by trying to recreate the conditions (the so-called "primordial soup") in which they believe it arose. As already mentioned, although they have not yet succeeded, that doesn't mean they never will. It may indeed turn out that life can spontaneously arise if all of the prerequisite conditions are met; that is, if there is a perfect blend of light, heat, water, air, minerals, etc. If it does turn out that way, atheists will have their long-held mantra that God is not *necessary* validated, or at least seemingly so. If it is proven that life can arise spontaneously, then it becomes much more plausible that the physical universe was never "created" at all; it has just always been here (in some form). Indeed, only if life cannot arise spontaneously by chance but requires an intelligent causal agent does God become *necessary*. Why? Because, to reiterate and emphasize a point already made—either God created *everything*, or else what we call God is himself nothing more than the most advanced life form that spontaneously arose by chance *within* the universe, in which case "God" would not really be *God* at all. If God is merely the first and most advanced life form, then He is bound by the laws of the physical universe rather than being above those laws. He is therefore limited. If He is limited, He is fallible. If He is fallible, He cannot meet the criteria for "God" laid out earlier in this study. Therefore, to make all else in *The Explanation* work, we must assume that God is the causal agent of *all* physical life, and not himself a product of chance.

The great difficulty in explaining the fifth day creation comes in first deciding how God went about forming "every living thing that moves" [NKJV], as the narrator puts it in verse 21. If the issue which we have already addressed of whether God created a single kind of plant life, three kinds, or thousands of kinds is problematic, the issue of whether God created a single kind of animal or more than one presents an even greater challenge. If the difference between simple algae and giant Redwood trees seems profound, how much more acute is the difference between single cell organisms and humans! Contrary to what secular evolutionists proclaim as fact, no amount of time could produce complex organisms like humans by chance. It simply defies all logic and credible statistical analysis to assume that time and chance will lead to that result. Everything else in the whole physical universe that we can observe or study shows the opposite result; it shows the law of entropy at work. Disorder, not order, results from time and chance. Things break down; they don't come together. It is mathematically and thus theoretically possible that life on earth is the exception, that it is the 1 in 10 to the millionth exponent chance occurrence. It is much easier to believe in an invisible supreme being, however, than to believe in such an exception. Besides, most secular scientists presume that earth is not the only place in the universe that harbors life. Life *must* be out there, they say; the universe is just too big, and it has too many billions of galaxies, each with billions of stars, for life not to have arisen by chance in more than one place. They add that, if only humans had the technology to peer far enough into the vast expanse of the universe, we would surely find another planet with life. That may be true, but it is neither verifiable nor falsifiable currently, and probably won't be in our lifetime. Thus, they take it on faith that someday they will be proven right, not unlike Fundamentalists (and some Quasi-Fundamentalists, too!) who believe it is only a matter of time until Jesus reappears and proves *them* (and me) right. The difference, of course, is that secular scientists take steps (build bigger telescopes, better satellites, more advanced communication equipment, etc.) to bring about the outcome they seek, whereas Fundamentalists and Quasi-Fundamentalists can only wait, watch, and pray for the return of Christ.

With that point established, the question must be asked: if God did create life anywhere else in the universe besides earth, would it violate the principles of Christian theology to the extent that irreparable damage would be done to the religion? The answer is no. The caveat, however, is that it definitely would violate a cardinal principle of theology to assume that God created more than one life form in his own "image." The whole Gospel *seems* to rest on the notion that God loved the species called "man" so much that He sent Jesus to die for it in order to redeem it. Just because that *seems* to be case, however, does not make it so. Cardinal principle or no cardinal principle, logic and rational thought allow for many speculations. What if humans are not the only creatures that have the "image" of God? What if God created another species in the universe, or many other species, in his own image? Would that require that they look like humans, or have other humanoid characteristics, such as cognitive skills and/or ability to fellowship with God or other heavenly beings? If they were to be in some ways like humans, must we automatically assume that they would become sinners, too, in need of a savior? If so, would Christ have had to die a whole separate death to redeem them? Surely Christ would not have died more than once, would He? And God certainly could not have had more than one "only begotten son" whom He could offer up for the sins of another species made in his image, else the Gospel becomes false or at least unreliable. But what if He had a daughter that He could offer up? What if He had a son who was "created" rather than "begotten" to offer up? Just because the Bible does not mention a daughter or a non-begotten son does not mean God does not have one, the other, or both. Or what if He has some other manifestation of himself that He could offer? If He manifests himself in the form of a Father, a Son, and a Holy Spirit, is it not possible that He might manifest himself in different ways to other extraterrestrial species? Again, just because the Bible doesn't mention them doesn't mean they don't exist.

With that said, unless and until life can be found on another planet, we must proceed under the assumption that life on earth is special, that it is unique, and that the only reason it exists is because God decided to put into a purely physical form what already existed in Heaven. Just as a great variety of angels exists in Heaven—cherubim, seraphim, messengers, warriors, etc.—so we assume that a widely diverse set of other creatures exists there as well. Just as humans are the physical creatures that have the image of God on earth, terrestrial animals have the image of some heavenly creature. Not only may there be a heavenly equivalent for every living earth creature, there also may be one for all the millions of extinct species. If so, and "variety is," as the old saying goes, "the spice of life" on earth, how much more true must that be in Heaven! If this hypothesis is true for animal life, there is no reason it should not be true for plant life as well. Therefore, if we could somehow take a peek into Heaven today, we would see the whole panoply of God's creation, of which only a fraction currently exists on earth. For believers who are impressed with the wonders of creation on earth, with all of the flaws and limitations, they will be astounded beyond words when they see the fullness of the creation in Heaven.

Getting back to the original point now, in verse 22 we see God "bless" the swimming and flying animals by telling them to be "fruitful and multiply" in order to fill up the waters and the sky. We must assume that the "blessing" came in the form of an instinctive sex drive; He blessed the animals with the will and ability to replicate their kind through a pleasurable experience. In so doing, He also endowed them with the ability to pass on genetic mutations by chance to their offspring, thus making evolution within their broad "kinds" possible. This concludes the fifth day of creation.

MILLENNIUM SIX

Verse 24 introduces Day 6 of the creation—the day in which "cattle," "beasts," "creeping things," and humans are all put on earth. Although this assortment looks strange at first glance, it seems clear that the narrator is referring to everything other than fish and birds here. Any and all animals that live on land (except humans) are included in this mix—mammals primarily, but also reptiles and amphibians. Whether these latter "types" or "kinds" of animals represent different stages in the evolutionary chain is the question upon which secular scientists and Old Earth Creationists will disagree. Science tries to show a progression from fish to land creatures with amphibians being the link. This progression seems to place reptiles one step beyond amphibians but one step short of both birds and mammals. However, as with *The Explanation*'s thesis about the evolution of all plants from three basic types or kinds, so too did the animals evolve from a few basic types, which God created separately: fish, birds, insects and other land-based creepy crawlies, amphibians, reptiles, and mammals. There is ample evidence that within each of these taxonomic groupings, mutations have occurred over time, such that it is rather easy to see how some species, such as coyotes, foxes, wolves, and domesticated dogs, for example, might have had a common ancestor. Evidence that members of any of these taxonomic groupings have evolved into something that resulted in a completely different taxonomic category, however, is at best sketchy and requires a leap of faith even larger than that required to believe in God. A prudent path for secular scientists to take would be to caution that evidence *suggests* evolution on this grand scale but does not yet *prove* it. Unfortunately, secularists dogmatically drive the narrative as if it is fact, and they do so to suit their preconception and sometimes to further the agenda of extirpating God and religion from the collective consciousness of humanity.

One advantage of believing in this "kinds" interpretation of the creation is that it allows for the rise and demise of species without destroying the nucleus of the DNA of each kind. To explain, we know that dinosaurs once roamed the earth but are now extinct. The seeds to make another dinosaur are still present in modern, living species, and could be resurrected (theoretically) by making some genetic modifications. In fact, it is theoretically possible that, given enough time, space, and human and material resources, we could reincarnate every single variety of animal that ever lived on earth at any time in the past by reverse engineering and genetic modification. This is important, because some atheists might object to the idea that a creator-God would ever allow some of his creation to go extinct. But logically, if humans could resurrect extinct species, then God certainly could, and these animals are thus not lost to him or Heaven despite being temporarily lost to us on earth at this time. (This whole topic makes us wonder whether God has a giant database of information somewhere in Heaven where He stores such things. Or maybe God, or one aspect of God, *is* a giant database of information!)

In verse 26, finally, we get to the creation of humans, or at least discussion of that creation (because the creation itself does not occur until verse 27). The thing that immediately jumps out at the reader here is the narrator's quoting of God as referring to himself in the plural rather than in the singular: "let us make man in our image. . ." Theologians have wrestled with this phrasing for centuries. It seems to contradict later passages in the Old Testament in which God wants his people to know that He is "one God," not multiple gods. The very basis for all western monotheism stems from the belief that there is just one God. So why would God refer to himself as more than one? The two most probable explanations are that 1) God had already created lots of heavenly beings in his own image before He created humans on earth, so He was talking to them; and 2) One manifestation of God was talking to the other(s), such as the Father talking to the Son or vice-versa. Either of these might be true, but most likely both are: the pre-carnate Jesus, known at that time as Yahweh (the bodily version of the spirit-God), was talking to his Father (the all-encompassing Elohim) *and* all the angels when He spoke in the plural.

That does not mean that the angelic beings played no role in the creation story. As previously mentioned, it is entirely possible that, once God brought physical matter into existence and produced life on earth, his celestial assistants performed the work of genetically modifying those life forms in order to speed the process of evolution along. If humans today can modify plants and animals through scientific research, why would we assume that more advanced beings could not or would not do the same? Clearly, God did not create these heavenly beings to sit around and do nothing, or to sing hymns and play harps on clouds all the time. He gave them a job of some kind while He unfolded his plan little-by-little. God did not tell them any more than He wanted them to know, however, so they were not privy to the plan in its entirety.

A second point in verse 26 that must be discussed is one that does not automatically jump out at the reader. The narrator says that the creator phrases his statement in two different ways: "let us make man . . ." 1) "in our image," and 2) "after our likeness. . ." Clearly God was saying two different things here. As already mentioned, "image" refers to outward, physical appearance, whereas "likeness" refers to internal and partly intangible characteristics. The point is that God wanted to make the human species to be like himself in both appearance and in character traits. And we know that Jesus, while living in his human body on earth 2,000 years ago, had both.

A third point in verse 26 is likewise routinely skimmed over by readers. The word "man" here seems to be plural rather than singular, because the narrator tells us that God says "let *them* have dominion . . ." The narrator does not say, "let us make *a* man" and "let *him* have dominion. . .," yet that is how we tend to read it. Why? Because we are projecting forward in our minds to what seems to be an individual "man" named "Adam," despite the fact that "Adam" is not mentioned until chapter 2 verse 18. This is important, because verse 26 lays out an explanation for God's intention *before* creating humans. His intention was to create a species that would "have dominion" over all the other animals He had already made. Why? Because any being that has the image and likeness of God must necessarily have dominion over other beings just as God has dominion over all of his creation. Here is the point, and it is vitally important to understanding the overall explanation: when we say God intended to "create a species that would 'have dominion'," that means God did not intend to create just *one* man and *one* woman and let them replicate all the rest of their kind through sexual reproduction; instead, God intended to create a LOT of humans individually as works of his own hands, just as He made Adam! Picture a master architect-engineer-builder who takes great satisfaction in creating original designs, constructing them, and bringing them to fruition, only in this case the master is designing and building new people. We should assume that this is how He created the angels. No one thinks that angels are products of sexual reproduction, do they? If not, why should we be shocked to think that God might have intended to create a whole host of humans the same way He created the angels? If that were the case, one might ask, then why did He ultimately allow replication through sexual reproduction? That question will be answered in the next chapter, and the answer will be fascinating!

It is extraordinarily important to notice that humans appear on earth on the same "day" that the higher order animals appear—the sixth day. Unlike what happened on Day 5, when God created fish and birds and told them to be fruitful and multiply, Day 6 sees the creation of all of the higher order animals *and* humans *before* instructing either of them to be fruitful and multiply; that is, before endowing them with an instinctive sex drive. Why is this important? Because it sets up the chronology of events in Genesis Chapter 2 in such a way as to render certain common hypotheses implausible and other alternative ones likely, as we will soon see.

Meanwhile, in verse 27, God actually creates humans, both male and female. It is interesting to note the similarities between humans and the higher order animals. Almost all higher order animals have bilateral symmetry, including two eyes, ears, nostrils, arms, and legs, but one brain, one heart, one mouth, one sex organ, etc. When He created humans, He used the same basic body type but modified it slightly. Clearly God was deliberate in his creation; therefore, we assume that He did not leave it to chance that one of the animals—such as the chimpanzee, which is one of the primates that shares the most similarities with humans—would evolve into another species that had his image and likeness. When God created humans, therefore, He did it as a separate and special action of creation. In Chapter 2 verses 7 and 19, we see that both humans and animals were created from either "the ground" or the "dust of the ground." But only of humans is it said that God "breathed" into the man the "breath of life," thus making him a "living being." Clearly, the animals were "living beings," as we commonly understand the term, too. So what gives? The answer is that the "breath of life" did not bring the human to physical life, but rather endowed them with the inner "likeness" of God. How can we know this? Because God did not breathe the breath of life into any other animal, yet they all came to physical life anyway—the animals all became animated. We will come back to these verses in Genesis Chapter 2 shortly, but first, we must finish Chapter 1.

Having created male and female humans in verse 27, God blesses them in verse 28, just as He had blessed the fish and the birds the "day" before, with the instruction to be fruitful and multiply; that is, He endowed them with an instinctive libido and with a desire for babies and children. (Just as eons of time may have passed between one verse and another at the beginning of the creation story, so it may be that a long, long time passed between verses 27 and 28.) Evidently, God gave most of the libido to the males of the species and most of the desire for babies to the females! Of course, each gender must possess a certain amount of either desire or at least toleration for the other endowment to make sexual reproduction a common occurrence. (We will see in Chapter 2 how this method of reproduction was NOT God's original, perfect plan for humans but was instead the result of sin.) God then instructs the humans to "fill the earth and subdue it . . ." In so doing, humans would then be able, ostensibly, to exercise "dominion" over all the other living terrestrial creatures and the earth itself, as God commands them in the second half of the verse. The natural assumption that most readers make is that the words "subdue" and "dominion" imply butchery and carnivorous use of the animals. That is, however, a false assumption. God does not command humans to kill and eat their fellow creatures but merely to bring them under authority; that is to domesticate them, to train them, to make use of them as living creatures. Cows were meant to be milked, for example, just as pigeons were meant to carry messages, and camels were meant to be beasts of burden. But animals were not originally meant to be eaten by humans or each other! This fact is abundantly clear by verses 29-30. God told the humans He had just created that He gave them "every herb" (vegetable) and all fruit-bearing trees (including bushes, vines, etc.) "for food." Then He told them that He gave "every green herb" to all the land, air, and sea creatures "for food." It does not say in either case that God gave them the right or the craving to kill and eat each other.

If it is indeed the case that God never intended humans and animals to eat each other, why do we do so now? The explanation is that carnivorous desire is a product of original sin. Only after the fall of humans from the perfect will of God, which is covered in Genesis Chapter 3, do animals receive the curse of becoming carnivores and omnivores. And make no mistake: living, conscious creatures killing and eating other living, conscious creatures is a curse! Secular scientists will no doubt protest this theological assumption, if not laugh it to scorn, because all of the evidence seems to suggest the impossibility of such a scenario. Evolution, they will say, brought carnivores and omnivores naturally onto the earth, just as it brought herbivores. The paleontological evidence for flesh-eating animals dates back millions of years. How can this point even be disputed then? Having no evidence to the contrary, we are faced with a theological dilemma. Either we can give in and say that verses 29 and 30 do not mean what they say, or we can argue and say, "Somehow, God made a way." It is in fact totally possible that the process of micro-evolution (mutation within species), which we have already explained herein, led to some creatures developing fangs, sharp teeth, claws, and a taste for meat and thirst for blood. It would be getting ahead of the story to say more about this topic for now, so we will return to it later.

Meanwhile, it should be noted in conclusion to our study of Genesis Chapter 1 that at the end of Day 6, God "saw everything that He had made, and indeed it was very good." He had not made anything that was *not* "very good." He had not made anything that would automatically lead to death, to unquenchable thirst, to insatiable hunger, or to some other unobtainable desire. Everything He made on earth was meant to be satisfying and easily acquired. It was an idyllic creation, full of life and abundance, not death and want. The introduction of death and want come in Chapter 3 in direct reference to the introduction of "sin." But before we get to that topic, we must plough through Genesis Chapter 2.

Genesis: The Explanation

What Really Happened
"In the Beginning"

2

THE EXPLANATION
GENESIS Chapter II

THE LABORATORY OF EDEN

Chapter 2 begins with an apparent contradiction. Chapter 1 had ended with the implication that the creation was finished as of the sixth day—that there was nothing left for God to do but rest and enjoy his work. Verse 1 of Chapter 2 confirms that thought and seems to make it a fact. Yet the next verse says that "on the seventh day, God ended His work . . . ," implying that the work was not finished *on* the sixth day after all but rather *on* the seventh. If so, the question immediately arises as to what hour of the seventh day? Was it at the stroke of midnight? Or perhaps one of the wee hours of the morning? Or was it mid-morning or some other time? In both verses 2 and 3, the narrator tells us that God rested either "in" or "on" the Sabbath day, but not that He rested for the full day necessarily, regardless of whether that full day amounted to 24 hours or some other measure of time. As mentioned earlier, this whole business of God measuring time is puzzling. If He existed before time, and if time was part of the creation of physical matter, why did He need to subject himself to it? In keeping track of it and making a deliberate decision to take a day off after 6-plus days of work, was He not indeed subjecting himself to it? If so, we must assume that He did so not for his own benefit but for the benefit of humans. Earth-bound humans would need an actual, universal clock to regulate their physical bodies and minds. So God established one, setting a pattern for its use by his own activities followed by rest.

> *1 Thus the heavens and the earth, and all the host of them, were finished. 2And on the seventh day God ended His work which He had done, and He rested on the seventh day from all His work which He had done. 3 Then God blessed the seventh day and sanctified it, because in it He rested from all His work which God had created and made. 4 This is the history of the heavens and the earth when they were created, in the day that the LORD God made the earth and the heavens, 5 before any plant of the field was in the earth and before any herb of the field had grown. For the LORD God had not caused it to rain on the earth, and there was no man to till the ground; 6 but a mist went up from the earth and watered the whole face of the ground.*
>
> *7 And the LORD God formed man of the dust of the ground, and breathed into his nostrils the breath of life; and man became a living being. 8 The LORD God planted a garden eastward in Eden, and there He put the man whom He had formed. 9 And out of the ground the LORD God made every tree grow that is pleasant to the sight and good for food. The tree of life was also in the midst of the garden, and the tree of the knowledge of good and evil. 10 Now a river went out*

of Eden to water the garden, and from there it parted and became four riverheads. 11 The name of the first is Pishon; it is the one which skirts the whole land of Havilah, where there is gold. 12 And the gold of that land is good. Bdellium and the onyx stone are there. 13 The name of the second river is Gihon; it is the one which goes around the whole land of Cush. 14 The name of the third river is Hiddekel; it is the one which goes toward the east of Assyria. The fourth river is the Euphrates. 15 Then the LORD God took the man and put him in the garden of Eden to tend and keep it.

16 And the LORD God commanded the man, saying, "Of every tree of the garden you may freely eat; 17 but of the tree of the knowledge of good and evil you shall not eat, for in the day that you eat of it you shall surely die." 18 And the LORD God said, "It is not good that man should be alone; I will make him a helper comparable to him." 19 Out of the ground the LORD God formed every beast of the field and every bird of the air, and brought them to Adam to see what he would call them. And whatever Adam called each living creature, that was its name. 20 So Adam gave names to all cattle, to the birds of the air, and to every beast of the field. But for Adam there was not found a helper comparable to him. 21 And the LORD God caused a deep sleep to fall on Adam, and he slept; and He took one of his ribs, and closed up the flesh in its place. 22 Then the rib which the LORD God had taken from man He made into a woman, and He brought her to the man. 23 And Adam said: "This is now bone of my bones and flesh of my flesh; She shall be called Woman, Because she was taken out of Man." 24 Therefore a man shall leave his father and mother and be joined to his wife, and they shall become one flesh. 25 And they were both naked, the man and his wife, and were not ashamed. [NKJV]

Chapter 2, as critics, skeptics, and unbelievers are quick to point out, repeats much of the same information already covered in Chapter 1. Some scholars believe that Genesis Chapter 2 is actually an alternative version of Chapter 1, narrated by a different writer who lived years later. They point out that the writing style in these two chapters is not consistent. Whether Genesis was written by one person, two, or a committee matters not, however, in trying to understand the book as a whole. Even if the writing style is not consistent, the overall story that is told in Genesis is consistent from one chapter to the next. Chapter 2 elaborates on or reframes some of the information in Chapter 1, such that the redundancy does not seem like overkill; it seems rather more like the elucidation of topics broached in the first chapter that went unexplained there. In Chapter 2, verses 4-7 show the first two examples of this, so let us look at them now.

In verses 4-6, the narrator explains that, prior to humans appearing on earth, no rain had ever fallen. Instead, a "mist" rose from the earth to water the land. It does not say that no vegetation of any kind had yet grown, only that no plant or herb had yet grown. The implication here is that none of the varieties of vegetation that produce food for humans had yet grown, but we must assume that grasses and weeds and any number of other kinds of vegetation had. This is all confusing unless we realize that these verses really seem to be mistranslated to place the emphasis on the wrong point. What is being described herein is actually farming; i.e., domesticating certain fruits and vegetables for human consumption. There was, needless to say, no farming prior to humans, but even after humans, there was no need for farming at first, because there were not seasons for rainfall and sunshine, flooding and drying out, planting and harvesting—*that* is the point of the verse. Instead, the earth was essentially a giant greenhouse that produced vegetation through a constant moisture and, we must assume, a regular, warm temperature. Chapter 2 does not follow the day-by-day creation timeline of Chapter 1. It seems to begin instead somewhere around Day 3 of the creation in verses 4-6 and then abruptly jump to Day 6 in verse 7. Because of this fact, it is consistent with the explanation already given to say there was plenty of time for mutation to occur in the plant and animal kingdoms, such that by the time humans appear in verse 7, there is a ready-made assortment of thousands upon thousands of different species of the various "kinds" of each.

Verse 4 contains several interesting points to consider. Depending on which English translation we use, it says "This is the *history* of the heavens and the earth..." or "These are the *generations* of the heavens and the earth." Although our current popular usage of the word "history" would fit in the context of the verse, the technical definition of it would not. History is technically the study of the past using written records. It requires that people living in the past leave a trail of evidence in the form of writings that a later generation can read or decipher. It is not an account of the past derived from archaeology, anthropology, paleontology, geology, or any other type of natural or social science. Information gathered from such sources is prehistoric, and it thus yields prehistory rather than history. Nor is history an account of the past derived from mythology or oral tradition; that is, written records made centuries or millennia after the time which they purport to tell about. It could be an account of the past given by divine revelation, of course, which is precisely what the book of Genesis seems to be, but that would do little to help us find the explanation we seek. If, however, we substitute the word generations for history here, we get a different picture of the past. We know that a generation is generically used as an ambiguous unit of time measurement which is roughly the equivalent of a human lifetime—some say it is 40 years, some 70, some 100, and some 120. There is no way to know for sure which is correct, or whether more than one may be correct, depending on the context in which it is used.

Beginning in verse 4, the narrator chooses to use the term the "Lord God" (Yahweh Elohim) instead of plain "God" (the Elohim) as the creator's name. Theologians have speculated that this may be because various parts of Genesis were written by different authors, and over a matter of centuries the various parts were assembled by editors into the current and final version of the book. According to this theory, starting in verse 4 of Chapter 2, we read the thoughts of a different author than who wrote the chapter and verses preceding it. Looking at Genesis from a purely Jewish perspective, this theory makes sense. Looking at it from a holistic Christian perspective, however, there is a better explanation. It is not that more than one author's hand crafted Genesis, but that the one author who wrote it all deliberately changed the *emphasis* from the Elohim—the heavenly Father God which exists apart from the physical universe at the beginning of the story to Yahweh—the physical manifestation of the Godhead who came into his creation and interacted with it as if He were part of it starting in verse 4 of Chapter 2. If so, what we are actually reading about throughout Chapter 1 is a brief summary of what the Father's *will* was and how it was accomplished, whereas in Chapter 2 we are reading about the person of the Godhead who was responsible for *carrying out* that will interactively. We get the distinct impression that this is the case by the verses that follow soon after verse 4. Prior to this verse, "God" (the Elohim) seems remote, mysterious, and mostly abstract. Thereafter, the "Lord God" (Yahweh Elohim) becomes increasingly personal and humanlike as the story unfolds—from planting a garden in verse 8 to interacting and conversing with Adam in verse 15 and beyond.

It would be wise to pause here for a moment to let that point germinate, for out of all the various puzzle pieces that serve as keys to solving the rest of the mystery, none is more important. Indeed, readers who grasp and embrace this point will see how many other pieces fit together. Readers who skim past it, or who do not take full mental stock of it, or who try but fail to understand it, will be frustrated and confounded as we move ahead in the story. So, going back to Chapter 1 and fitting in this piece, we get: "In the beginning, the Elohim created the physical universe and all that lies within it, and the Elohim did so while existing apart from it and outside of it. The Elohim set all the laws of nature in motion—gravity, magnetism, mass, the speed of light, the colors of the spectrum, the elements, etc., yet the Elohim was not subject to those laws, for the Elohim is supernatural (above and beyond nature)." As long as God stayed/stays above and beyond the physical universe He created, He was/is the Elohim. Once God enters his creation here in Chapter 2, however, that manifestation becomes known as Yahweh (the Lord) for the rest of the Old Testament. "The Lord" is actually more of a title or description of his role in the drama than it is a proper name. The actual proper name of "the Lord" is Jesus ("Yeshua" in Hebrew), although people living before the physical birth of Jesus, the son of Mary, could not have known that. But we Christians certainly know it now, so by extension we know that Yahweh is Jesus and Jesus is

Yahweh, because there is only one Lord in existence and therefore only one Lord mentioned throughout the Bible, and it is He. Thus, to avoid confusion, for the rest of *The Explanation*, whenever referring to the "Lord God" (Yahweh Elohim), we will call him simply "The Lord." Readers are encouraged to visualize "the Lord" of the Old Testament (Yahweh) and "the Lord" of the New Testament (Jesus) as one and the same in every case—different manifestations, but the same being. It is not unlike how the pre-resurrection Jesus and the post-resurrection Jesus are the same being, yet they have certain dissimilar attributes, the one having been an earth-bound suffering servant, the other the eternal, glorified King of Kings.

To recap and return to the story, beginning in Genesis 2:4, the Lord interacts with his creation as if He was/is part of it. In so doing, He does not replace or supersede the Elohim, and the Elohim therefore does not cease to be God. The Elohim remains the spirit manifestation of the Godhead always, simultaneous to the Lord playing the central role on the inside of the creation as God physically manifest—tangible, or made knowable to created beings. To say that the Lord proceeded/proceeds forth from the Elohim is the same as the New Testament saying that Jesus proceeded/proceeds forth from the Father. Yahweh is God made tangible to beings locked inside the physical universe; Jesus is likewise God made tangible. Yahweh is Jesus; Jesus is Yahweh—one is a name, the other a title, but they are merely different manifestations of the same being. This being, "the Lord," makes appearances on the inside of creation for the purpose of interacting with humans, who are his special creation because we are the only ones made in his image with the likeness of the Elohim.

MAN OF DIRT

In verse 7, the creation of humans is discussed again, as it was in Chapter 1, but with a twist. Here the Lord took the "dust of the ground" and formed it into a "man." Before going any further, we should notice that this dust was derived from the same planet that had just been described as having a mist that "watered the whole face of the ground." In the absence of some good explanation, this seems to be a contradiction. Did God somehow dry the ground in order to make dust? If so, did He dry the whole earth temporarily? Or did He dry a single spot? Perhaps He scooped up a pile of mud just the right size for making a man and dried that? This is impossible to know for certain, but a reasonable explanation would be that the phrasing of "the whole face of the ground" does not mean every square inch of the earth's land surface. We must assume that the earth's topography featured hills and valleys, mountains and seashores, etc., and that although the mist bubbled up from underneath the ground, it did not automatically make every surface point wet. We can also assume that during the heat of the day the sun evaporated any moisture on the surface, leaving pools only in low-lying areas and producing dust, dirt, and sand in higher areas.

There is, of course, a whole different possibility for how to interpret this passage. The "dust of the ground" could simply refer to the elements, minerals, and chemicals found on or within the earth's surface. In that sense, God created man from the same materials which He had used to make the other animal life forms. That, in fact, seems to be the point of the narrator saying man was made from dust. It was a statement designed simply to remind the reader of man's lowly origins, not to teach the biological or geological science behind his creation. The Lord took "something" into his hands and turned it into a human—that is the point. Whether that something was literal dust or merely a strategic assortment of items plucked from the periodic table, the Lord formed it into a being in his own image with the Elohim's likeness.

As an interesting aside, in the New Testament, in John Chapter 9 we see Jesus heal a blind man by daubing mud on his eyes, and we might do well to suppose that He did so as a proof that He was in fact the same being as created humans in Genesis by using dirt. The interesting thing about this miracle in John 9 is that it follows immediately after John 8:58 where Jesus reveals that, "Before Abraham was, I am." The Pharisees reacted to this statement by trying to stone him. So by healing the blind man with mud, Jesus is essentially confirming that He was and is who He claimed to be—a manifestation of the Lord Yahweh of the Old Testament.

Regardless of how the specifics of the creation of the first human occurred, the Lord then breathed life into him. The human then became a living "soul" or "being." Interestingly, the Bible says that God created all the other animals prior to man and gave them life, but it doesn't say He specifically breathed life into them or made them into souls. This is the main distinction about humans that made them have the likeness of God. They had the very breath of God in them, which means a consciousness that contains a conscience. While all higher animals have some type of consciousness, we assume that none has a conscience. If some seem to have a primitive conscience, there is as yet no way to measure it. Perhaps science will find a way someday, but for now we can safely say that no animal has a conscience anywhere near as sophisticated as humans have. Most of what they do, they do by instinct. Some of what higher order animals do, they do by experiential learning. Regardless, they serve only their own self-interest and are not capable of true altruism or philanthropy. Moreover, whereas some higher kinds of animals, such as dogs, dolphins, and apes can learn to do new things like humans do, they cannot originate new ideas, and they cannot invent or improve upon existing technology, secular evolutionary theories notwithstanding. More importantly, they cannot reflect upon their actions and become philosophical about them. They can remember that to touch fire will cause pain, so they know not to touch fire again. But they cannot learn anything about the cause of fire. Again, science may someday progress to the point where human trainers are able to teach primates how to gather sticks, strike a match, start a fire, and cook food, but it hasn't happened yet. Even if it did, would that somehow prove that those higher life forms have a soul or a conscience? No.

What exactly is the "soul," anyway? Is it synonymous with the conscience, as suggested herein? Is it something that survives the death of the body? Is it measureable? Is it quantifiable? Can it be made visible? Or audible? No one knows for sure, of course, but those of us with faith believe that it does survive the death of the body. There have been enough well-documented cases of near death experiences in which unexplainable things have happened—such as a patient lying on a gurney in a hospital, flatlining, rising up out of himself or herself, and seeing the lifeless body beneath them, etc., to satisfy those who *want* to believe that something exists beyond this physical realm. The soul, it seems, is normally trapped in the body until the body dies, upon which the soul is released into the great invisible beyond. If so, where was the soul before the person was born? Some speculate that it existed in Heaven prior to being inserted into a particular body at a pre-appointed time. According to this belief, God has been warehousing unborn souls for eons until each one's appointed time and body arrives on earth. He then dispatches them, as if on a spiritual conveyor belt, into those bodies. While that indeed may be the case, there is a reason to think it is not, but rather to believe that the soul is created inside the body. In other words, it does not exist anywhere at all prior to being developed with the body in the mother's womb. The womb serves as an incubator or scientific laboratory which nurtures the soul along with the body. Why this must be so will become evident shortly.

Meanwhile, moving on to verse 8, the narrator tells us that the Lord planted a garden on the eastern side of Eden, that He put Adam there, and that a couple of important trees were located there. Before discussing Adam and the trees, we must first examine the geography of Eden. The precise location of Eden is not identified here, but scholars have always speculated about it, mostly using verses 10-14 as the basis. Those verses tell us that a single river ran *from* Eden to the outside world, where it then divided into 4 parts. Three of the four parts do not have modern equivalents that can be ascertained with absolute certainty, but it has one part that can be—the Euphrates River. We know the Euphrates starts in the Caucasus Mountains where Turkey is today and flows in a southerly direction into the Persian Gulf. Some scholars speculate that, if there ever was a place called Eden at all, the narrator either got his facts wrong or the topography of the Middle East was so different back then that the Euphrates flowed in the opposite direction! Wherever it was located specifically, it was broadly within the area that is today the Middle East. Whether that was Mesopotamia or some stretch of land that has since been covered by the sea, it was obviously somewhere in the vicinity of where the Euphrates is today.

The most reasonable presumption to make is that, prior to the great flood of Genesis Chapter 7, the topography of the whole earth was different. In fact, there may have been only one giant land mass—one super-continent originally, and the splitting of the land into continents may have been what caused the deluge of Noah's time. Be that as it may, it is clear that many low-lying areas were surely covered by the flood, never to return as dry land. Practically the whole Mediterranean Sea was once dry ground, for instance. The Mediterranean region is thus a possibility for where Eden might have been located. Two of the prime candidates that scholars have identified, however, are the upper Persian Gulf, just below the Tigris-Euphrates delta, and the Black Sea, just beyond the Strait of Bosporus. Theologically and historically, the Persian Gulf would seem to be a better choice, but geologically and archaeologically, the Black Sea might be. Both of these possibilities are commonly proposed because there is evidence that at some point in the past each was either above sea level or at least not covered by water, and archaeology indicates that people lived there.

Beyond the general location of Eden, we know nothing about it. We do not know its size, for instance. We do not know if it was an area of, say, a few hundred acres or many thousands of square miles. We also cannot know whether it was the place from whence the Lord took the dust to create Adam, or if so, whether it was the place where He performed that creation. We tend to assume it was where Adam came from, not just where the Lord placed him afterward, because it is a natural assumption to make. The text doesn't say that, however. A more substantive hypothesis is that, regardless of where Adam was created, it did not take place within the garden that lay within the broader region called Eden, because verse 8 says the Lord *put* him in the garden. Verse 15 repeats that point, adding that the Lord *took* and *put* him there. Incorporating the ancient alien theory into this story, we can imagine that the Lord had a laboratory somewhere, be it in space or inside a spaceship sitting on the earth, and there He formed Adam from the elements He had taken from earth. He then moved Adam into Eden and deposited him within the garden there. What this garden actually was is subject to just as much speculation as anything else in this story. It probably was not what we think of as a garden in the conventional sense but rather an ecological laboratory on earth within which Adam was assigned to work as a cultivator of some kind. We will return to the topic of what it meant to cultivate the Garden of Eden shortly.

Verse 9 introduces an extremely important new topic into the story: the "tree of life" and "the tree of the knowledge of good and evil," but it does so almost as a peripheral point that is no big deal. Consequently, it seems somewhat out of place here. The narrator lumps these "trees" in with all the other trees that bear edible fruit, as if they are merely two separate species of fruit trees. One can imagine the narrator expanding on this verse, saying, "God created the tree of apples, the tree of oranges, the tree of coconuts, the tree of life, the tree of almonds, the tree of olives, the tree of the knowledge of good and evil, the tree of pecans, the tree of pears, the tree of figs, etc. . ." These two special trees are presented here, in other words, so nonchalantly that the casual reader will likely skim right past this verse without much thought. Scholars, however, have puzzled over this curious wording for centuries. Many have speculated that these two "trees" were not actual trees at all, but rather were called that allegorically because there was no closer equivalent that the narrator could use to describe them. We will come back to this topic and explore it in some detail shortly, but first let us go back to Adam and make some illuminating observations about him.

Nowhere in this account is it said that Adam was created in the form of a grown man. We assume that he was grown because our English word "man" is used to describe him. "Human" would be a better word. If we substitute "human" for "man" throughout the story, we can easily see how Adam may have been created in the form of a child. This is important because it helps clarify later parts of the story, as we will see. It raises some unanswerable questions, though: if Adam was a child, at what age-level was he created? Seven? Ten? Twelve? If he was created at any age level other than 0-1, how was his chronological age determined in Genesis 5:5, which says he lived to be 930 years old? He certainly would not have been created as a new-born baby, an infant, or a toddler ... or would he? Well, maybe. If the Lord was able to build the human creature, wiring it complete with DNA that controlled physical and mental growth and development, He was undoubtedly able to feed and care for a new-born or rear a toddler. Just because it is difficult for us to visualize such a scenario does not make it an impossibility.

Let us go a step further, in fact, and hypothesize that God created Adam as an embryo in a sort of Petrie dish in a laboratory. Through some kind of artificial womb, He nurtured the growing human until he was ready to be unsheathed from his cocoon. Thereupon, Adam needed all of the same kind of care that any new-born baby needs. Rather than having a mother to suckle, the Lord (by use of the angels perhaps) fed the babe with what we would call today formula from a bottle. In time, the infant became a toddler, able to eat solid food, crawl, and speak in baby-talk. At that point, perhaps the Lord moved him into the Garden of Eden. There, Adam, growing up around the Lord and the angels, learned to walk, talk, and do for himself. As a child, he, like all of us as children, experienced life as a wonderland of new things to see, hear, touch, taste, smell, and learn about on a daily basis. As he grew and was able to digest and incorporate new information, he received the education necessary to be able to carry out the job that God had assigned him in verse 15—to dress and keep the garden, meaning ostensibly to farm it or cultivate it. Like many of the other verses in Genesis, this one must be considered allegorically. This particular type of cultivation was more like what a scientist would do today than what a farmer would do. It was experimental, environmental, and educational. What all it entailed, we cannot know. It may have involved advancing earthly industry and technology as well as growing and hybridizing plants and selectively breeding animals. Regardless of what all it involved, when the time was right, the young man Adam was finally prepared mentally to be put to the task without "adult" (angelic) supervision. Thereupon, in verses 16 and 17 the Lord gave him the one and only rule he was to observe: do not eat from the tree of life or the tree of the knowledge of good and evil. Before delving any deeper into that topic, we must set the stage by discussing an adjoining issue.

Verse 18 broaches the subject of Adam's aloneness in the garden. Reading casually, the verse would seem out of place here, as if it should come after verse 19 and the first half of 20 (which discuss the naming of the animals). If, however, we apply the above-mentioned notion of Adam having long enjoyed the company of angels or the Lord himself in the garden while he was growing up and being educated before being left alone without supervision to do his job, then it makes sense that this verse comes first. While he had been under the tutelage of the Lord and/or angels, Adam had no opportunity to experience loneliness or to feel that something was missing from his life. Upon being turned loose to do his job without supervision, however, it was only a matter of time until that began to change. We assume that the aloneness he experienced came in the form of loneliness for human companionship, but that is not necessarily so. In fact, it probably was not so at first, because the Lord actually sought a "helper" for him among the animals. That fact makes verse 18 and the second half of verse 20 two of the most bizarre sounding verses in the whole Bible. Did God really try to find a companion for Adam that was outside of his own species? The only way this makes sense is to reject completely and utterly the notion that the type of companionship God was seeking for Adam was sexual. It most certainly was *not* sexual! It was more along the lines of the type of companionship that man's best friend, the dog, provides many of us today. But even the most ardent dog lover will tell you, if he or she is sane, that there is no substitute for human companionship. Did God not know that in advance of searching through the various animals for Adam's mate? Surely He did. So why the story takes this strange turn is a mystery.

In an attempt to clear up the mystery, let us entertain a possibility or two. Let us assume that the Lord knew there would be no helper suitable for Adam among the animals, but He chose to let Adam find that out first-hand for himself. He could have simply told Adam as much, but He wanted to show him. That way, when He introduced Adam to a female of his own species, Adam would fully understand and appreciate the significance of this gift. Adam would see the stark difference between his own species and the nearest-in-appearance of other species, such as apes. There would be no comparison, and Adam would instinctively recognize the contrast between the one as sacred and the other as profane. From then on, he would never crave the companionship of any animal in the same way he would crave that of the female human. Then when God had introduced Adam to all the various species, let him name them, and instructed him to have dominion over them, the instruction would be as clear as it could possibly be. So He brought the animals to Adam one species at a time and let him name them but also interact with them. He knew that Adam would try to communicate with animals and to school them in the same way that the Lord and the angels had been communicating with and schooling him. He knew that Adam would then quickly realize the futility of this endeavor and look to his maker as if to say, "Is this all you've got for me, Lord?"

Moving on in the story now, when the time was right and Adam was ready, the Lord brought the various species of animals to him for naming. Adam named them in his own language according to whatever characteristic(s) about each one stuck out to him. Whatever he called a "giraffe" in his own language, for example, probably meant big mammal with a long neck or something equally descriptive. Whatever he called a "snake" probably meant something along the lines of curvy slithering reptile. Although the Bible does not say that Adam was given the job of naming plants, too, it does not say that he didn't name them, either. He probably did name them. The skeptic will say that Adam could not have possibly named all the different animals and plants on earth because new species are still being discovered today! But the Bible does not say he named every living thing on earth. It implies that he named whatever he saw in the Garden of Eden only, and that may have been a fraction of all there was on earth or, as a laboratory, it may have been a factory that pumped out various mutations. In other words, the earth might have been overspread by certain "kinds" (prototype species or taxonomic groupings) from the garden, and it was only those kinds that Adam named.

How long did this whole process of Adam's growing, learning, and naming things take? There is of course no way to know. If we take the genealogy in Genesis literally, the first few generations lived for hundreds of years each; and there is good reason to take it literally: God created man with the intention of his living forever, like the angels. Man's disobedience (sin) led to his death, however, as the Bible makes clear. Even so, it was not a mystical, spiritual principle that caused death, as if God simply issued a decree that the *punishment* for sin would be death. Instead, God put a natural law in place in which disobedience brought *consequences*, and the main consequence was genetic degradation over time, as we will see in the next chapter. For now, considering the vast life spans of humans in those early, pre-flood generations, it may very well be that it took fifty years or more for a child to reach puberty. If so, as the average life span decreased over time, so did the age at which humans reached puberty. This is merely conjecture, but it helps explain some otherwise hard-to-explain things in Genesis. Regardless of how long this growth process took for the first human, at some point he finished the job of naming the things the Lord assigned him to name. Meanwhile, he was still mastering the knowledge of how to "keep and dress" the garden and, inevitably, he eventually mastered that, too. At that point, he then found himself, as we all do when we reach a certain stage of life, feeling more than just the aloneness previously described; he now felt an *emptiness* that he had never experienced before. Naturally, he did not know what to make of it. Whether he complained about it to the Lord, or whether he just took on a sad expression, the Lord recognized it for what it was: it was the emptiness that can be filled only by the companionship of a fellow human, and perhaps especially that of the opposite sex. And having let Adam find out for himself that there would be no satisfactory companion among the animals, the time and conditions were now right to put the plan in motion to create a female human.

Here we must now go back to verses 16 and 17 and make a preliminary explanation of the "tree of the knowledge of good and evil" which God commanded Adam not to "eat of." The full explanation will come shortly, and it will make sense at that time. For now, suffice it to say that this so-called "tree" was not an actual *tree* but rather the thing that was capable of producing fruit from Adam's own body—his genitalia. The Lord told him essentially not to touch or handle it. The verb "eat" in these verses would be better translated "partake" or "make use" of. We understand the reason: doing so would cause arousal, and once aroused, nothing would cure that problem except a sexual release (orgasm/ejaculation). God knew that the moment such a thing happened, Adam would lose his innocence and thus sow the seeds of his own demise as a physical creature. There would be no going back. There would be no more childlike existence for this pure soul. Once he discovered the physical pleasure of sex, his whole existence from then on would be devoted to satisfying that craving, just as food satisfies hunger and water satisfies thirst.

The man would not stop until the craving was satisfied. The Lord did not want that for Adam. He wanted his human creation to live free of such a burden and to live forever. If so, why would God put this "tree" literally within arm's reach of the man? The first part of the answer is that, we must remember, at the time God gave the prohibition of partaking of this "tree," Adam did not know what he was missing. Having never experienced the feeling of pleasure brought by sex, it was actually easy for him to accept the prohibition. We will offer the second part of the answer shortly, but first let us finish tracing the story to the creation of "Woman."

ADAM'S RIB

To take the Genesis account of this whole topic at face value would be to think that the Lord discovered a design flaw in Adam when He realized the human was lonely. Clearly, God does not make mistakes, so any time in Genesis a plain reading of the text seems to imply that God made a mistake, we know immediately that either the translation is worded incorrectly or our understanding of the translation is faulty. The Lord knew when He programmed the DNA of Adam that he would reach this point in his growth. It did not catch him by surprise. He was prepared to respond to it. Thus we see the "Woman" now introduced into the story. Much-ado-about-nothing has been made through the centuries regarding how the Lord caused "a deep sleep to fall over Adam," whereupon He took a "rib" from Adam and made his female companion. Intending a reading that is sympathetic to the plight of the supposedly-lowly second sex/gender, some preachers have dogmatically stated that God chose a rib from the man's "side" because woman was meant to be by man's "side" eternally, as opposed to beneath him, behind him, or above him. The fact is, God could have taken any part of Adam—toenail, spinal disk, skull fragment, or whatever, and accomplished his purpose. Why? Because when we look at this story from a purely scientific point of view, it is obvious that the narrator is trying to convey the message that the Lord anesthetized Adam and performed a minor surgery on him in which He took a DNA sample from a bone. He then reengineered the X and Y chromosomes to make a female version of a human, which He presented to Adam as his "wife" (verses 24-25), and which would be better translated here as soul mate, best friend, or most intimate companion.

One obvious question that arises in light of this scenario is, "why didn't God make male and female humans simultaneously instead of one at a time?" The question is a sensible one considering that God apparently made all other animals in male-female pairs. To answer, it would be helpful first to ascertain what the purpose of the male-female relationship was/is in the animal kingdom. Animals seem to pair up in the wild primarily for instinctive procreation rather than companionship. In the higher order animals it is certainly noticeable that they desire socialization within their species beyond the brief act of copulation, but it does not seem to require members of the opposite sex necessarily. Males of some species seem prone to hang around with other males, while females hang around with each other or their offspring until those offspring are weaned. In cases where a particular alpha male hangs around with his females, it seems to be for possessive, territorial purposes rather than companionship. It has little if anything to do with love. There are, of course, all kinds of variations on the socialization patterns of animals. Some species, such as doves, which pair up with a single mate for life, seem close to our human concept of love, but their actions are merely instinctive, not chosen through free will. And therein lay the great difference. In the case of humans, God gave male and female alike a soul—a conscience with free will in all things, including love and fidelity.

To continue laying the foundation for answering the question previously posed, we must take it for granted that the Lord made Adam in a male body, not a female body, and not an androgynous body. Why? Because the Lord himself had taken on that form (male) from the beginning. He had also given that form to the angels, who were/are male in physical appearance, at least in all cases where the Bible mentions them, but were designed to be asexual in nature. Adam was created in his and their image. Just because Adam had the "plumbing," so to speak, to be able to perform sexual functions, did not mean the Lord intended him to use his genitalia for that purpose any more than He intended the angels to use theirs for that purpose. Why give him the plumbing then? For one reason, because the main body part that distinguishes the male anatomy from the female, and vice-versa, is used for urination as well as copulation. Clearly the human body was created for eating, drinking, and processing food and water through the bowels, so that was the one undeniable purpose of the plumbing. Using it for sex, just because it *could* be used that way, was not the Lord's intention.

To state the obvious, this all seems like a cruel joke. Why would an omnipotent God create humans, animals, and angels alike with body parts that can be used interchangeably for sex and urination? Surely He didn't have to. An omnipotent God could have made two separate parts. Or He could have made humans and angels completely incapable of sexual activity if He had wanted to. So why didn't He? The answer is because that would have eliminated free will. Some may argue that, biologically, we don't really have free will when it comes to desiring sex; the desire is thrust upon us, and all we can do is fight it off or submit to it. That is true today, but it is only true because of original sin. Prior to the fall, which will be explained shortly, Adam did not experience sexual desire. Instead, Adam's desire was to please the Lord, to make him proud, and to be rewarded for obedience and a job well done. The Lord did not create Adam, therefore, to have sexual relations like the animals, but He did create him with the physical capability to do so. And the Lord knew that by giving Adam a female companion, the temptation would inevitably arise. Again, if it seems that the Lord set these humans up for failure, remember that there was no other way for him to give humans true free will. Where there is no temptation or desire, there is no freedom of choice. It is a harsh reality, but it is *the* reality. But as we shall soon see, it was not the man's desire for sex that led to the first act of disobedience; it was the woman's desire for something else.

Now we are ready to answer the question, "Why didn't God make male and female humans simultaneously instead of one at a time?" The answer is mainly because He wanted to give each one a different responsibility. He gave Adam the responsibility of learning from him and the angels and then passing on this knowledge to the woman. He gave the woman the responsibility of learning primarily from the man, and simultaneously gave her the responsibility of being the man's helper or assistant in the garden. This is all evident when Genesis is interpreted in light of the New Testament teachings on the respective roles of women and men. Paul spends a considerable amount of time trying to untangle the mess that ancient societies had made of gender roles. His exposition in I Corinthians Chapter 11 is exemplary in this regard.

There in verse 3 he says that the head of *every* man is Christ. This began with Adam as he learned at the feet of the Lord himself in the garden (with angels playing a tangential role). It goes on to say that the head of the woman is the man. That, too, began with Adam and his female companion in the garden. The English language does not have an exact equivalent, so "head" here must be translated to mean "leader," "mentor," and/or "role model." The point is that Adam was given the assignment of being the woman's teacher. The male had a big head start on the female in terms of knowledge, in other words, and he was charged accordingly with the task of getting her up to speed. This in no way implies that all males were meant to be permanently "over" the females, just that the first male, created in the image of God, had to lead the first female, who was created from his body. As we will see in the next chapter, all humans thereafter would be created from the body of both male and female, making them more equal than what some Fundamentalists believe because of their interpretation of certain New Testament passages, such as the one cited above.

To conclude this chapter now, in verse 23, Adam is introduced to his special new companion, whom he welcomes undoubtedly with great joy. He names her. Interestingly, the narrator does not say that the Lord instructed him to do that. Adam simply knew to do it. It is what he had done with the animals; it is one of the things he had been trained to do. He calls her "Woman" here, not "Eve." Woman means "out of man." He does not call her "Eve" until Chapter 3 verse 20, after the fall. We will discuss that name in the next chapter. The narrator then inserts a parenthetical commentary in verse 24 which foretells the whole history of the human race after the fall. We will save discussion of that verse for a later chapter as well. Verse 25 concludes the creation story by noting that both Adam and the woman were naked but were not ashamed. This is the lynchpin verse of the whole chapter. It explains in summation why the "tree of the knowledge of good and evil" was, and could only have been, sex.

3

THE EXPLANATION
GENESIS Chapter III

TO DIE FOR

Just as the Bible does not specify whether Adam was an adult when created, so it does not say whether his female companion was either. We can speculate about it, though. Would it be more plausible that this second human was created as a fully-formed adult or as something else? Although we know that God is omnipotent and can do anything in any way He chooses, we also have established that He seems to choose to follow the "rules" of the physical universe that He himself created. If He, through his bodily manifestation called "the Lord," created Adam using a physical, earthly, scientific process, why should we think He created Adam's female companion any differently? If He created her from the DNA of the man, then He would have, it seems, grown her the same way He grew Adam. Whether that was in a sort of Petrie dish or some other controlled laboratory environment, we can assume that she was not formed fully grown. She would have been put through the same basic educational regime as Adam, but in her case Adam would have been her primary instructor instead of the Lord or the angels—a vitally important point that we will return to shortly. At what age level or height-and-weight she had attained when the Lord brought her to Adam, we cannot know. Did Adam have to wait for years after her creation to meet her? Or did the Lord speed up the process somehow to prevent a long wait? Did Adam even know the Lord was creating a companion for him, or did the Lord just surprise him with this gift? Either way, it seems likely that the Lord used a normal earthly time frame, which would have helped Adam learn patience. And it would have given Adam time to move beyond merely having enough knowledge to mentor the woman; he would have perhaps been able to master the things the Lord had shown him, to begin learning from his own experiences and experimentation, and to pass this experiential knowledge along to his female helper.

Once Adam met the woman, he named her. Then the two of them lived in the Garden (laboratory) of Eden for an unspecified period of time, with Adam serving as instructor and his female companion as apprentice, before the fateful day of the arrival of the "serpent" and the infamous temptation incident. They witnessed the two "trees" being utilized—one by the animals and the other by God and the angels. Each tree did the same thing—created life, but each accomplished it in a different way. Even though both trees created life, God called only one of them "the tree of life." The other He called, strangely it seems at first glance, "the tree of the knowledge of good and evil." The man and woman did not understand the concept of good and evil initially because they existed in a state of childlike innocence. They actually may have still been children, if we consider that it may have taken 50 years or more to reach puberty, as mentioned in the previous chapter. To repeat, the Bible does not say Adam and his female companion were created as adults. We always assume that to be the case because tradition would have it that way. Yet there is nothing in the Bible that even hints at what relative age, size, or maturity level they were created. To digress, it also does not say that they looked white with blonde hair, or black with kinky hair, or swarthy with curly hair. We must assume that they had attributes in their DNA that could branch out in their progeny with great variety over time. Regardless, they were childlike in their innocence, so they did not understand what either good or evil was at first, but neither did they question what they were. Or if they did, it is not recorded in the Bible. When the so-called serpent arrived, however, it induced them into questioning what good and evil were, as well as questioning other things that God had told them. The serpent, in fact, provoked them to become skeptical of God's authority and cynical about God's motives. And as we will see, when the serpent tempted the woman with the forbidden fruit from the tree of the knowledge of good and evil, he made it look good enough to die for! The woman thus made a choice that brought a series of consequences that humans have been dealing with, and a curse that all living creatures have suffered under, ever since. Let us continue following the story now, so we can understand these life-producing "trees" and determine what this "serpent" was that brought on the fall of humans.

> 1 Now the serpent was more cunning than any beast of the field which the LORD God had made. And he said to the woman, "Has God indeed said, 'You shall not eat of every tree of the garden'?" 2 And the woman

said to the serpent, "We may eat the fruit of the trees of the garden; 3 but of the fruit of the tree which is in the midst of the garden, God has said, 'You shall not eat it, nor shall you touch it, lest you die.'" 4 Then the serpent said to the woman, "You will not surely die. 5 For God knows that in the day you eat of it your eyes will be opened, and you will be like God, knowing good and evil." 6 So when the woman saw that the tree was good for food, that it was pleasant to the eyes, and a tree desirable to make one wise, she took of its fruit and ate. She also gave to her husband with her, and he ate. 7 Then the eyes of both of them were opened, and they knew that they were naked; and they sewed fig leaves together and made themselves coverings.

8 And they heard the sound of the LORD God walking in the garden in the cool of the day, and Adam and his wife hid themselves from the presence of the LORD God among the trees of the garden. 9 Then the LORD God called to Adam and said to him, "Where are you?" 10 So he said, "I heard Your voice in the garden, and I was afraid because I was naked; and I hid myself." 11 And He said, "Who told you that you were naked? Have you eaten from the tree of which I commanded you that you should not eat?" 12 Then the man said, "The woman whom You gave to be with me, she gave me of the tree, and I ate." 13 And the LORD God said to the woman, "What is this you have done?" The woman said, "The serpent deceived me, and I ate." 14 So the LORD God said to the serpent: "Because you have done this, you are cursed more than all cattle, and more than every beast of the field; on your belly you shall go, and you shall eat dust all the days of your life. 15 And I will put enmity between you and the woman, and between your seed and her seed; he shall bruise your head, and you shall bruise his heel." 16 To the woman He said: "I will greatly multiply your sorrow and your conception; in pain you shall bring forth children; your desire shall be for your husband, and he shall rule over you." 17 Then to Adam He said, "Because you have heeded the voice of your wife, and have eaten from the tree of which I commanded you, saying, 'You shall not eat of it': Cursed is the ground for your sake; in toil you shall eat of it all the days of your life. 18 Both thorns and thistles it shall bring forth for you, and you shall eat the herb of the field. 19 In the sweat of your face you shall eat bread til you return to the ground, for out of it you were taken; for dust you are, and to dust you shall return."

20 And Adam called his wife's name Eve, because she was the mother of all living. 21 Also for Adam and his wife the LORD God made tunics of skin, and clothed them. 22 Then the LORD God said, "Behold, the man has become like one of Us, to know good and evil. And now, lest he put out his hand and take also of the tree of life, and eat, and live forever"— 23 therefore the LORD God sent him out of the garden of Eden to till the ground from which he was taken. 24 So He drove out the man; and He placed cherubim at the east of the garden of Eden, and a flaming sword which turned every way, to guard the way to the tree of life. [NKJV]

THE CHOICE

The temptation story begins in verse 1 of Chapter 3 with the serpent, which was more subtle, sly, or cunning than "any beast of the field which the Lord God had made," asking a question rather than making a statement, presenting a point of fact, leveling a criticism or accusation at God, or framing a case against him. This agent of temptation (whoever or whatever it was) gave the appearance of presenting the woman with an opportunity to teach him something, to give him knowledge on an issue he seemed genuinely curious about.

After she answered in verses 2 and 3 with the information that Adam had given her, the tempter mischievously began to correct her, doing so in small, polite, unthreatening increments in verses 4 and 5. Basically, the tempter said the tree of the so-called "knowledge of good and evil" was not what God said it was, or at least not what the woman understood it to be. The tempter admitted that the tree was indeed the source of some important knowledge, but it had nothing to do with "good" and "evil." Instead, knowledge that would make her and Adam like God and the angels was what it was really all about. This was a half-truth, which this agent of temptation knew would be much more believable than an outright lie. How long the conversation between the tempter and the woman lasted, we cannot know. Nor can we know what all was said, but the gist of it surely resembles the description given above.

Contrary to a literal reading of the passage, this agent of temptation did not come to the woman in the form of a talking snake. One clue that leads to this conclusion is that the narrator says this serpent was different from "any beast of the field," which should be interpreted as "all other animals." He did not say different from any *other* wild animal, but different from *any*, meaning all. It was therefore a unique creature, whatever it was. Tradition would have it that the mind or spirit of Satan possessed the body of the serpent, making it talk and giving it its slyness and deceitfulness. It is certainly possible that a Satan type being—a fallen angel that could take on different shapes and appearances—may have indeed come to the woman and first planted an idea in her mind, just as the Bible story implies. Whether that was the case or not, a Satanic (anti-God) *thought* certainly came to her, as such thoughts typically come to us today—in the form of a question, a doubt, or a feeling that something doesn't add up. So, what didn't add up in the woman's mind? The answer: this whole business of life—how it is created and how it reproduces. Perhaps she could see that the Lord and the angels had created life through the "tree of life" (a scientific process in a laboratory environment), or perhaps she had never seen that process and thus did not know about that method of creation. Either way, she could certainly see how, once life was created initially, it could reproduce on its own through a completely different process. For animals, the process was sex.

Following the trail of breadcrumbs in this story, we can only arrive at one conclusion: the so-called "tree of the knowledge of good and evil" was sex. Why? Because sex was the *literal* "tree of life" for all other animals besides humans. Adam and the woman had seen it. They had watched animals copulating, not understanding what it was for, but in their innocence not questioning it either. For an unspecified period of time—maybe years or decades, they had seen various species of animals engaging in sex acts. Having, like children, not experienced it themselves, they could imagine neither the pleasure to be had in sex nor the consequences to follow from it, so they did not know to desire it. They had also watched animal offspring being born, but in their innocence did not immediately make the causal connection between sex and childbirth. Over time, they made that connection, but it started with the woman, not the man.

There was a reason—a very important reason, the so-called serpent approached the woman rather than the man. When the Bible speaks of women being the weaker of the two sexes (I Peter 3:7), it is not referring to some kind of physical or mental inferiority. It is instead saying that women have "a" particular weakness that men do not have. They feel an unexplainable (and sometimes uncontrollable) need to have children. The Lord made it instinctive in all females throughout the animal kingdom, including humans. So it was specifically the woman who first made the connection between sex and childbirth, but only after some outside force (serpent) nudged her mind in that direction. While Adam was busy doing as the Lord had instructed him in keeping and dressing the garden, not letting himself question his job or the rules of living in the garden, the woman began noticing that she had the same basic body parts as the female animals that gave birth. And she noticed that Adam had the same parts as the males of the animal kingdom. At some point in time, she became aware of the male genitalia, which perhaps only coincidentally resembled a serpent, and realized it was the thing she needed in order to produce life. (Some scholars have suggested that Adam's phallus was in fact the "serpent" that "spoke" to the woman, and there was no other being present in the garden). Perhaps she had noticed that, upon waking in the morning, Adam had an erection, as many men are prone to have because of the bladder filling up during sleep. Her mind was thus open to a new thought which neither the Lord, the angels, nor Adam had put there. Until she acted on it, however, her eyes were not yet opened. Nor were the man's. The man, still in a childlike state of innocence, was actually the weaker of the two sexes in his ability to grasp this type of new information not given him by God. The woman was a quicker study so to speak, and because of her desire for a child, she had a proverbial chink in her armor that an outside force could exploit. In that sense, she was the first to be tempted with disobeying God's command not to eat of that tree, or, in other words, not to partake of that life-producing activity. The "fruit" of that tree would be a baby, and to the woman the mere thought of it was so alluring as to be simply to die for.

Good and evil? What did any of this have to do with good and evil? The female human didn't know, but she knew what life was, she saw how it was made, and she wanted it. She didn't crave sex to satisfy a sex drive; this wasn't about libido or lust. She wanted it to satisfy the craving for what the female animals all had—a baby, a child. To her, sex was merely a means to an end. To her, therefore, this all seemed to be about making a choice, not succumbing to a temptation. To get the thing she wanted, however, she had to entice Adam. She had to pass along to him the revelation of the serpent. She said to him, in essence, "If you and I do what the animals do, we can create life like they do!" We cannot know precisely what she said to Adam, but it surely came across about like this made-up quote. If it sounds overly simplistic, remember that the woman was approaching the issue like a child setting out on a small adventure. So as simple as it sounds, this scenario is the most plausible real-life explanation of what happened in the garden. Divorced of the mysticism of a literal, talking snake and a mature but naive woman, this is a version of the story that actually makes sense.

Whether there was a physical snake-like being in the story depends on whether the Lord allowed the fallen angels to enter the garden. He may have, because otherwise how would the free will decision already described ever come about? Conversely, He may not have except in this one case in order to create in the humans or introduce to them free will, because Adam and his companion were theretofore basically just automatons. He may have allowed Satan to take the form or guise of a snake-like being instead of some other creature because the snake stood out among the animals in terms of being stealthy, and the Lord was tricking Satan into thinking he was sneaking into the garden unbeknownst to its angelic gatekeepers. Or, for an alternative explanation, the Lord may not have allowed a physical serpent-like creature into the garden at all, because it may be that how the temptation of the woman transpired was no different than how the temptation of any of us occurs today—by arising in the mind. Some temptations may seem to be spontaneous, arising from thin air, but most are actually caused by a physical stimulus of some kind—something we see, hear, touch, taste, or smell. The woman's watching of the animals as they engaged in reproductive activity may have been the only stimulus necessary. Even if there was a snake-like, satanic physical creature that started it all, Adam, upon feeling the pleasure resulting from the sensation of touch through intimate relations, was inherently tempted by sex from then on, just as all of us are after experiencing it for the first time. After the first time, there is no need for a physical being to whisper lustful thoughts in our ear or to show us a visual stimulus. The mere memory of the feeling produced by sex is enough to spur us to action. It is a force that overtakes us which can hardly be stopped. The reason this is true for us today is because the animalistic instinct of reproduction through sex was encoded in our DNA by the actions of the first man and woman. That's why the "Satan" character throughout the Bible does not need to be physically present in each of our lives today to get us to sin. It comes naturally to us through our DNA. But the question remains. Was there a physical snake-like creature and/or a shape-shifting Satan character in the garden? We know there was *something* or *someone* that acted as an anti-God influence on the woman, because the Lord addresses it as a living being in verse 14.

Let us assume there was a physical Satan character present in the garden which took on the form of a 4-legged serpent-like being. What might it have looked like? What might it actually have been? One possibility is that it had the appearance of a mythical dragon, having not only legs but also wings, yet otherwise having a body much like a snake. Picture it as possessing the capability to speak to the female human in her own language. And to make this plausible, consider it as a space craft that had the form of a dragon, piloted by fallen angels that could speak the language of God and the first humans. Then it makes much more sense how this story of the serpent could be realistic. Another possibility is that the term "serpent" actually means the same thing as our modern English word "dinosaur." In other words, the serpent might merely have been a large reptilian creature of some kind that no longer exists on earth. If that is the case, it would explain a great deal, which we will discuss shortly. Whether a physical reptilian-looking creature of some kind spoke to the woman, an extraterrestrial entity or craft did it, or whether it was merely the woman's own thought that caused her to disobey the Lord, the spirit of the anti-God character was present in the garden then just as it is ever present in each of our consciences today. It was and is the spirit of disobedience, the spirit of pride, the spirit of selfishness, and it was all passed down through the ages in our DNA thanks to one act of original sin.

Consider this: why should there not be a Satanic, anti-God character in this story? Christians believe there was a Satan in the story of the temptation of Christ (Matthew chapter 4), and Jesus spoke to Simon Peter at one time, saying, "Get behind me, Satan" (Matthew 16:23). There is such a character mentioned in several other places in the New

Testament as well, including in some detail in the book of Revelation. There was such a character in the story of Job, too, and Satan is mentioned in the Old Testament in 1 Chronicles 21:1 and Zechariah 3:2 as well. So who was or is this Satan character? Is he (or *it*, as the case may be) the same being throughout the Bible, or might there be various satans? The book of Revelation indicates that he is one being, and he is the arch enemy of God and Christ. Whether it was *that* exact Satan who made all those appearances throughout the Bible cannot be known. But if it were some other being, that other being was ultimately under the influence and direction of *the* Satan. Just as when Jesus spoke to Simon Peter calling him Satan, it seems clear that He was really speaking to the "spirit" of disobedience to God's will that had manifested in Simon Peter's words, not saying that his number-one disciple had literally become *the* Satan. But that there was and is a being in existence whose mission has always been to oppose God and to convince humans to oppose God as well is obvious, for without evil there can be no way to recognize good. So without a Satan (anti-God figure), there can be no way to appreciate God.

Getting back to the story and thus to the overarching explanation, Satan's goal from the beginning of the creation of man was to reduce God's special creation—humans—to the level of animals. Why? Because he knew these beings were created in the image of the Lord and the "likeness" of the Elohim, and he knew God would not consider them as reflecting his likeness (even though they would always have his physical image) after they had corrupted themselves with animalistic sex. He knew that once the man experienced the pleasure of sex and the woman experienced the satisfaction of creating and nurturing a life through her own body, there would be no going back. He also knew that the Lord would be true to his word about allowing these humans to die a natural, physical death—just like the animals did—if they disobeyed him. Did Satan not know that humans had been given immortal souls the same as himself and the angels? Or did he assume that the spirit would die if the physical body died? Either way, Satan was not (and still is not) an all-knowing being. He did not know of the plan of redemption in advance any more than humans knew it, until after Christ finished his work on the cross, whereupon the plan was made manifest to him just as it was to humans. At the time he just knew his goal was to destroy God's special creation, or at least to hurt it, and this was a great way to do it, probably the best way to do it, and maybe the only way to do it. Sadly, it worked, and humans have been dealing with the consequences of the fall ever since.

THE CONSEQUENCES

In verse 7 of Chapter 3, the narrator begins telling the story of these consequences: "Then the eyes of both of them were opened, and they knew that they *were* naked..." The second phrase here shows undoubtedly that the first phrase cannot be taken literally but must be understood as an idiom. Adam and the woman's actual *eyes* were not closed, but their *minds* certainly had been. Once they committed the act of disobedience, their minds and/or consciences were automatically turned on, just as the serpent had foretold! They suddenly realized they had done "wrong" simultaneous to recognizing what nakedness was. Exactly how much time elapsed between the commission of this original sin and their mental recognition of good and evil is impossible to know, but we should assume that it was nearly instantaneous with the climax of the act. We can all relate to how this occurred, because it occurs to each of us at some point in our childhood—the point at which we become "aware" of sex. Even before actually engaging in sex, just becoming aware of it immediately pushes us out of the stage of innocence into which we are all born. Once pushed out, we can never go back. That's what happened to Adam and the woman as well.

They now understood the difference between good and evil, but they also quickly noticed that, although the bad thing they did had brought a pleasure they could not have even imagined, that pleasure was fleeting. Moreover, it was followed by a feeling of guilt and shame that lasted much longer. In fact, it wouldn't go away! And thus the infernal, internal, and seemingly eternal, tug-o-war was on. From that point on, the man and woman and all of their progeny would be engaged in a never-ending quest to obtain this temporary pleasure again and again. They would also be in a perpetual state of confusion about the rapturous emotions that accompanied the act, as well as a permanent state of frustration about the emotional let-down that followed it. Meanwhile, the immediate problem for Adam and the woman was the guilt and shame that accompanied the sudden recognition of their nakedness. They did not know what to do about this new-found problem; there had been no training given by the Lord or the angels for dealing with such a scenario. They did know, however, what clothing was and how it was used, because they had seen it on the Lord and the angels. They had never needed to know what they wore it for, but now they understood. They would have to improvise, and improvise fast. How much time elapsed between the discovery of their nakedness and their decision to sew fig leaves together and make themselves coverings (or "aprons" as the old KJV puts it) is impossible to know, but the wording seems to suggest that it occurred that same day and perhaps even within the hour. Whatever the time frame, they drew upon their knowledge of keeping and dressing the garden to come up with the plan for how they could use leaves from a fig tree to cover themselves. Sewing these large, fuzzy leaves together could be accomplished by using fibers from the stems of the leaves themselves or any number of different kinds of vines. The humans would have been familiar with all of the various plant possibilities that grew in their garden habitat.

Verse 8 tells us that Adam and his wife "heard" the "Lord God" walking in the garden in the "cool of the day," while verse 10 tells us that it was specifically the Lord's "voice" that they heard. There is much to unpack here. First, they apparently *heard* the Lord coming before they *saw* him. The Lord in this verse was clearly a physical being, the bodily manifestation of the Elohim. He could not have been a spirit, because a spirit would not have made a sound, unless it was a sound like unto wind perhaps. We get the sense from the context that this being was physically present because He was "walking," not hovering, not gliding, and not appearing out of thin air. Moreover, this being apparently waited until a time of day when it was not hot to do this walking. Whether that means early morning or late afternoon, we cannot know, but the narrator frames the story in such a way as to lead us to think it was late afternoon/early evening. Either way, only a physical being would need to be concerned about the heat; a spirit would be impervious to a property of the physical universe such as heat. The verse goes on to say that the man and woman hid themselves from the Lord, as if they knew He had physical eyes and a limited ability to know their whereabouts.

If so, it would undoubtedly be because they had interacted with him long enough to recognize that He had the same type of physical body and mind as they did, that they themselves were in fact something akin to replicas of him and the angels.

Verse 9 seems to confirm this view by saying that the Lord called out to Adam, "Where are you?" Some theologians may say that the Lord already knew where they were, but He intentionally got down on their human level, pretending that He didn't, in order to make the circumstances more meaningful to them. Whether He was omniscient or not, either way, it seems clear that He really did want them to reveal themselves. He did not want to have to expose them; He wanted them to give themselves up voluntarily. This would have been step one in getting them to take responsibility for their actions. He could have found them, uncovered them, and yelled out, "Aha! Caught you!" He could have proceeded to berate them for being bad. But He did not. He chose to speak to them with a certain type of respect, like a CEO might talk to his Board of Directors, not like a drill sergeant to a raw recruit at boot camp, and not like a strict schoolmarm to a first grader who can't sit still. Yet Adam's response in verse 10 comes across as if he was ashamed and fearful, like a child answering a grown-up, anticipating the worst in terms of an angry, perhaps violent, reaction from the Lord. Indeed, he said as much: ". . . I was afraid because I was naked . . ."

The Lord answered Adam with two more questions: "Who told you that you were naked?" and "Have you eaten from the tree of which I commanded you that you should not eat?" The first question seems to imply that the Lord knew Adam could not figure out that he was naked on his own; he could have known about his nakedness only by being told.

But that implication contradicts what we have already said about the consequence of sex being a natural, immediate awakening of the conscience. Therefore, the first question must have been rhetorical. The second question that the Lord asked Adam implies that it was perfectly logical to link nakedness with eating from the tree of the knowledge of good and evil. That was, apparently, the first and only thing the Lord immediately raised as a possibility (keeping in mind, of course, that if the Lord was omniscient, He didn't have to ask for his own benefit, but He verbalized his thoughts for the benefit of the humans and the serpent).

It is important to note that in verse 9, the Lord did not call Adam and the woman out together, but instead called out Adam alone. He addressed Adam first, as if talking to the senior member of his staff. He, in fact, did not address the woman until after Adam pointed the finger at her in verse 12. Would the Lord have even addressed her at all if Adam had not blamed her for their disobedience? Is it possible that He might have punished Adam alone and left the disciplining of the woman to Adam's discretion? We cannot know for certain, but we might do well to assume that He would have addressed the woman sooner or later and notified her of the coming consequences of her actions. If it had been *punishment* that the Lord wanted, He may have left it to Adam to determine. Instead, what He did was not issue a punishment against the woman, but rather simply notify her of the natural *consequences* of engaging in a sex act like the animals did: she would now have a baby in the same way the animals have them—painfully, dangerously, and bloodily—rather than the way God intended (through the tree of life).

Likewise, the Lord did not address the serpent until the woman pointed the finger at it in verse 13. Once He did address it in verse 14, the language sounds as though He were talking to an actual snake, which adds even more mystery to the identity of the creature than was already there—and plenty was already there! What can it mean that the Lord told the serpent that it would be cursed beyond all other animals, that it would lick the dust and have the human's heel ever on its head, while it would be reduced to striking merely at that heel from below? A literal reading of the text seems to yield an obvious answer—one that has been taught to children in Sunday School class for generations—the snake once had legs like most other land animals, but God's curse removed them and left the snake to slither along on its belly. But surely there is more to this story than that. What about the snake's voice? Clearly it was a talker; it had talked the woman into disobeying the number one rule of life in the Garden of Eden. What about the snake's brain? It was, we must assume from the text, at least as advanced, if not more advanced, than the woman's, for it was able to persuade and deceive her. Did God take away the snake's voice and highly developed brain along with its legs? It would be irrational to think that is the proper explanation.

Applying a common Christian theological concept, we get something that makes more sense. Consider that the serpent represents an advanced life-form which has more knowledge than the humans; call it Satan or whatever. When God curses it, He issues a *punishment* against it, unlike what He did in merely notifying the woman of the *consequences* of her actions. God strips this creature of its ability to walk among the humans in their sight, indwell another living creature, and talk in an audible voice. He consigns the creature to either a spiritual dimension or else to an extraterrestrial existence, where it cannot interact visibly or vocally with humans, except perhaps in rare cases by special permission, as happened with Jesus in the wilderness. The reason the Elohim allowed Satan to tempt Jesus is twofold: 1) because Jesus was born without the corrupted DNA of all pure humans, so He did not inherit the sin nature that the rest of us have. Therefore, to get Jesus to sin required Satan to actually, physically show up and talk to him, just as He did with the woman in the Garden of Eden. 2) because the Lord Yahweh and Jesus were/are one and the same, and He wanted to demonstrate his son's superiority over Satan, which Jesus accomplished by refusing not one but three temptations, in order to prove his qualifications to redeem humans later through his atoning sacrifice. Whereas the woman in the garden was presented with one temptation and immediately fell for it, Jesus endured forty days and three different types of temptation without succumbing.

As for Satan having invisible, inaudible influence over other humans, the Lord allowed and still allows it. Before explaining how and why, it would be helpful to say how and why *not*. Not the way that virtually all theologians for at least the past 2,000 years have assumed—by tempting us individually as he did the woman in the garden and Jesus in the wilderness, albeit via telepathic messages sent to each of our subconscious minds. It is rather impossible for most of us to visualize this without seeing the childish caricature of a little red devil with horns and pitchfork sitting on one shoulder whispering bad things into one ear, while a little white angel with a halo and wings sits on the other shoulder countering the temptation in the other ear. Despite its childishness, that cartoon depiction of how temptation works does *seem* to reflect the real-life mental tug-o-war that takes place within us all on a daily basis accurately. Even so, it is inaccurate. It is equally inaccurate to think that Satan sends us mass delusions collectively. The reason this is not possible is because Satan is not capable of omniscience and omnipresence like God. He cannot read minds, and he cannot be in more than one place at a time. The former would be necessary in order for him to tempt any of us individually, and the latter would be necessary in order for him to tempt more than one human at a time. Does he then have a host of demons whom he sends out one-by-one to tempt individual humans? Yes and no; to say only yes is too simplistic, but to say only no is misleading. In ancient times, a host of fallen angels pretending to be "gods" certainly interacted with humans, convincing them to believe and do things that would result in bad consequences (not punishments). The Lord allowed it, as we will see in the next few chapters, in order to produce maximum free will among humans. Once started, this Satanic influence would carry over from generation to generation, reproducing the same bad consequences, although with a multiplying effect as the population increased. Sickness, disease, war, famine, poverty, slavery, and various kinds of depravity are some of those consequences.

The real explanation for how temptation works, though, is actually just as simple, but otherwise completely different: temptation would be passed from one human to another via genetics. Once the woman and Adam had been deceived, disobedient, dishonest, and thus defeated spiritually, they would pass those traits on to their offspring just the same as they would pass on all other physical, emotional, and psychological traits they possessed. Satan knew that, or at least suspected it, which made it worth his effort to enter the garden and interact just that once with humans in the flesh. If so, he must have known he would get caught, right? Maybe not.

There is a good reason to think otherwise. Consider that, according to common Christian theology, Satan was formerly called Lucifer, the highest ranking being in Heaven besides any manifestation of the Elohim. While in that position, he "thought" or "believed" he could defeat God and his forces or at least be equal to them, like a co-God or something. If he were omniscient, he would have known the outcome: that he could not possibly win; so surely he wouldn't have made the mistake. If that is not the correct explanation, then it must be that God did not give Lucifer free will, but rather predetermined his fate by planting the seed of disobedience in his mind. Yet, that conflicts with the standard Christian belief that Satan is the "father" of lies (meaning of sin or disobedience in general, regardless of what form it takes). Yet again, if Lucifer/Satan is the "father" of *anything*, wouldn't that make him a "creator" like God? If so, one of two things must follow—either Satan created one and only one thing in his entire existence (call it "evil" or whatever), or Satan is capable of creating other things, too, and perhaps even of creating anything at any time, just like God.

Most Christians will have a hard time accepting the latter possibility, yet it would potentially explain some mysterious things about our physical world, such as where mosquitoes, viruses, and other scourges which seemingly have no "good" function came from. Despite the potential use of this possibility as an explanation for certain hard-to-explain things, most Christians are probably correct to reject this idea. A more likely explanation is that, while in Heaven, Lucifer could create, but once he created evil, he was cast out and stripped of his ability to create. Thereafter, he has been forced to work within the limitations of God's creation, just like all other angelic beings. Having already created evil, he was capable of spreading it to any other beings whom he could deceive, whether celestial or terrestrial. That means God allowed him to take the one thing he had created with him as he left Heaven, perhaps as an eternal reminder of his foolishness, or perhaps because God had a larger plan—the plan of redemption of the humans in mind before He even created them. Maybe it was both. Either way, Lucifer/Satan, having already deceived those angels which were deceivable, moved on to the only other target—the terrestrial beings.

Assuming this explanation to be accurate so far, we must now jump another theological hurdle. Was it possible that Satan could deceive earthly animals other than humans?—like, say, snakes? The only way that could be possible is if the animal(s) had a brain that was high-functioning enough to make moral choices. This seems like another theological bridge too far. We do know, however, that Satan's minions—the demons—can "possess" animals, as when Jesus allowed a group of demons to enter a herd of swine in Matthew 8:28-34 (although we should remember that the herd ran violently into the water and drown, thus signifying the possibility that animals do not have the same capacity for hosting demonic spirits as humans). If that is the case, we must assume that Satan himself could possess a snake or other reptilian creature. Such a possession could potentially account for the serpent in the garden. To those who would respond that the serpent would have needed vocal cords of some kind to communicate with the woman, we can recall that God spoke through a donkey in Numbers Chapter 22. If that passage of scripture is to be believed, then should we not assume that other equally unnatural things might have happened within the animal kingdom over the millennia? Moreover, consider that in this story of Balaam and the donkey, the animal was able to see an angel which was invisible to the man. There can be no doubt that many animal species have abilities that humans do not have. Examples include the migratory instincts of birds and the impending-danger perception of dogs. So, when Genesis 3:1 says the serpent had certain characteristics that no other animal had, there are precedents in the animal kingdom for that to be possible.

With that said, there is another possibility. Even though Satan was stripped of his ability to create by fiat such that he could never again make something from nothing, he was not stripped of his intelligence, which was undoubtedly extraordinary. His intelligence probably exceeded that of all other created beings, even the faithful archangels. By any standard, therefore, he would be a super-genius. As such, his scientific knowledge would be almost limitless. If so, would he not have been able to modify things within God's creation (assuming, of course, that God allowed it)? And if so, might not he have modified a reptilian creature to give it the ability to speak? He would not have needed to modify the whole species, just one specimen. Bringing this whole discussion full circle now, it is possible that Satan actually thought he could get away with it. And in a sense, he may have been right. Consider that the Lord cursed the serpent, not the initiator of the evil within the serpent—Satan. Did Satan receive any punishment for his part in the deception of the woman? Not directly. Christians, of course, tend to read the verse about heel and head (Genesis 3:15) as prophetic of Christ defeating Satan in the end, and that interpretation may be right.

Speaking of this prophetic take on scripture, it is well worth our time to digress to that topic for a moment. Consider the possibility that, prior to the creation of humans, God had no intention of punishing Lucifer beyond stripping him of his power and casting him out of Heaven. His punishment would be eternal separation from all that was desirable; he would be made to live in outer darkness where he would wander for all eternity. Once God created the earth and the various life forms on earth, however, this being now had somewhere to go and something to do that gave his otherwise hopeless and meaningless existence a purpose. Upon arriving on earth, he immediately set out to sabotage God's creation, especially his most beloved creation, the humans. He becomes the enemy of mankind by his own choice, and as such, he then becomes "Satan" rather than "Lucifer." At that instant, he has made his second mistake, and now God must punish him again. This time God chooses to punish him not merely with continued separation and meaningless wandering but with eternal torment. Although He does not reveal his plan to Satan, his plan of redemption is already underway.

Satan can grasp parts of the plan, but not all of it. He can't grasp it all at once, either. He picks up bits and pieces of it throughout human history, just like the various prophets of God do. As each one is given another piece to the puzzle, Satan learns it, too, and uses it to try to solve the puzzle. He can never do it, though, not through the thousands of years leading up to the time of Christ on earth. But over time, he gets closer. He sets in motion, for instance, the mythologies of pagan religions in which a "son" of a "god" is born, dies, and is resurrected. Also, once the angel announces the divine impregnation of Mary, he sets out to kill the baby, but God will not allow it. Even when Christ arrives in the flesh, Satan still doesn't fully understand the mystery of the plan of redemption. He senses that something very important is happening; he knows who Jesus is. He tempts him accordingly, for forty days in the wilderness (Matthew Chapter 4), hoping to work the same deception on him that he worked on the humans in the garden. But alas! He fails. He tempts him at other times, too, as in the case when he provoked Christ into rebuking Peter, saying, "Get behind me, Satan" (Matthew 16:24). He always fails. He cannot get Jesus to fall for any deception. However, when he succeeds at deceiving others to put Christ up on the cross, he thinks he has won. Even though he correctly anticipates Jesus' resurrection, he does not know and cannot know the outcome of the story. Not until Christ enters Hell for three days and releases the souls of Satan's human prisoners does he begin to understand. But not until John receives and writes down the Revelation on the island of Patmos does he, like the rest of us, learn the extent of his fate. By that point it had become too late to alter his course. Finishing what he had started—taking as many human souls with him into eternal damnation as possible—gave him and still gives him his only purpose in existence. Just as he had/has no more power to create, he also has no power to kill or destroy spirits or souls. Therefore, he could not put himself out of his misery by committing suicide of the spirit. His fate is sealed, and that is why he continues to this day in his quest to harm humans. And to think—it all started in the garden. It turns out that he, the super-genius, more than the woman, was the naive one in this story!

Getting back to the main topic now, the Lord addressed all three participants in the drama of disobedience in the garden in reverse order. He addressed the serpent first, followed by the woman, and then Adam. His curse upon the serpent was to become the permanent enemy of the woman and her "seed." This means that even if Satan intended his temptation of the humans to be a one-time event, God would lock him into a perpetual, dysfunctional union with the humans until such time as He finished his plan of redemption—although, as already mentioned, Satan didn't know what the plan was or how long it would take. Some translations say that God would put "enmity" between the serpent, the woman, and her seed, meaning they would be enemies, each trying to destroy or at least hurt the other. If we take as fact here the hypothesis that the serpent actually represents the whole animal kingdom, it makes a lot more sense. Humans and animals would now forever be locked into a struggle for survival. The animals would seek to attack the humans and would sometimes do so successfully, but mostly the humans would dominate and kill the animals. Even if we take the alternate hypothesis that the serpent represents dinosaurs/reptiles, it makes more sense than the orthodox explanation. Humans would ultimately come to dominate the planet while the dinosaurs would become extinct. Although this sounds like a mutual curse for all participants, God's curse upon the serpent was, of course, more severe than the pronouncement of consequences to Adam and the woman, because He informed the serpent that the best he would be able to do in this dysfunctional relationship was strike the humans at the very lowest point of their physical bodies—the heel. The humans, however, would be able to strike the serpent at his very highest point—the head. This was, therefore, by no means a mutual curse. God in essence informed Satan that he would be doomed to lose this struggle. God thus consigned him to unending frustration. Satan would seek to destroy the woman and her seed, but he would be repeatedly defeated, and this would go on (we know now) for thousands upon thousands of years on earth, not to mention into spiritual eternity.

Many theologians believe that the "seed" of the woman refers to Christ (in the singular), rather than to all humans who were to ever be born of women (in the plural). This interpretation may well be accurate; there is a good reason to think so. It gives a certain symmetry to the plan of redemption which it otherwise would not have. Even so, it is not necessary to interpret Genesis 3:15 this way in order for the story to make sense or for the plan of redemption to be carried out. If we interpret "seed" to mean all the people ever to be born on earth, it actually helps verse 16 make more sense. Here in verse 16, the Lord addresses the woman and notifies her of her fate: she will now, as we've already discussed, bear children just like the animals do. Beyond that, however, she will also be forever subservient to her husband, meaning females of the homo sapiens species will be in a permanent *position* of inferiority to the males of the species—and the emphasis is on "position." Inferior *status* does not equate to inferior *personhood*. As already noted, the first woman, the ancestral mother of us all, was actually smarter, or at least more perceptive, than Adam in certain ways. And so it has been ever since. Women and men as a whole have complementary but different mental skill sets, and the ones that women typically possess in abundance which men do not possess make them smarter than men in those areas. Those who say the Lord created the woman as an inferior being are therefore wrong. The Lord created the woman to help the man, it is true, but He did not create her to be under the dominion of man like the other animals were to be, or to be bought and sold like property, or to be otherwise controlled or abused by men either physically or psychologically. Yet, in this pronouncement of consequences, God notifies the woman that such will be the fate of her female children for all generations to come, until He should say otherwise. Those of us who have the good fortune to live in modern times in the more advanced nations of the world, which not coincidentally tend to be the nations with Christian backgrounds, may be inclined to respond negatively to that assertion, because we look around us and see females enjoying a high level of equality with men. Were we to look into the traditionally non-Christian nations, however, with their poverty and lack of education and modernization, we would see a different picture. Well over half of all the females alive on the planet today are in this inferior position, not in a position of relative equality. Therefore, the pronouncement of consequences still applies to more women than not today. If so, imagine how much worse it must have been in ages past!

Verse 16 also says that the woman's "desire" would be for her husband. In light of the interpretation we just discussed, the translation of that phrase would be better rendered, "The woman's desire shall be for *a* husband." In other words, the continuing consequences for all generations of females born on earth would be that they would need a man in their lives in order to attain a good quality of life for themselves. They would need him for procreation and protection, if nothing else. Moreover, they would have this desire for children imbedded into the fabric of their being, just as the first woman had it in the garden. It would be passed on from generation to generation in their genetic code. It would not be a choice, therefore, but an unrelenting *need* which could only be satisfied by a man. Of course, we can find exceptions throughout history. In the Christian age, we see that there have been thousands upon thousands of nuns, for example. In modern secular culture, we see lesbians and heterosexual "liberated" women as exceptions. All-in-all, however, these are merely the exceptions that prove the rule. For every one woman who takes an alternative lifestyle, there are easily a thousand who don't.

Moving on to verses 17-19 now, we finally come to the Lord's addressing Adam. He begins by telling him why a curse is about to come not on himself but rather on the "ground" (translation: *whole earth*): "Because you have heeded the voice of your wife . . ." The Lord says, in effect, "I told you not to partake of that one thing—just one thing. I gave you free reign over everything in the garden, and all I required in exchange was for you not to touch that one forbidden fruit. And yet that wasn't good enough for you. So, let's see how you like this new arrangement I'm about to make for you. How about this: from now on, you can have that one thing. You can have it all you want. Just copulate and copulate to your heart's content. Go ahead. Have a big ol' time doing it. But in exchange, I will take away your easy life in this garden. Now you will have to work for everything you once received freely—securing food in the wild will become your new job. It won't be fun, and it won't be easy. The very earth underneath your feet will supply it, just as it did in the garden, but it won't grow of its own accord, and it will only grow amid a tangle of thorns, thistles, and weeds. You will spend your days foraging for food, and you will spend your nights worrying about where you will find the next day's meal. The pleasant task I gave you of keeping and dressing the garden will now be replaced by physically exhausting tasks such as plowing, planting, harvesting, shepherding, milking, churning, preparing, and cooking. And this will be your lot day after day, year after year, and generation after generation for all time to come until I say otherwise. You will be charged with providing for your wife and for the offspring the two of you will bring into this world because you couldn't say no that one thing."

THE CURSE

It is important to note that in verse 18, the Lord says "you shall eat of the herb of the field," but nowhere does He say that Adam and his family would be forced to eat their fellow creatures, the animals. That may be implied as part of the curse, however, because in verse 21, the Lord made clothes for the humans from animal skins. The narrator does not tell us whether Adam and "Eve" (who was dubbed that by Adam in verse 20, meaning "mother") watched as the Lord killed and skinned the animals. If they did, we cannot know whether they were forced to watch or merely allowed to watch. If they were forced to watch, it could have been for punishment, education, or both. If punishment, what better way to drive the point home that disobedience to God has real and serious consequences than to show Adam and Eve what their sin led directly to: the slaughter of their fellow, innocent creatures! Imagine the shock that must have seized the minds and the remorse that must have flooded the soul of the humans—a confusing, sickening, tormenting feeling of blood libel that none of us will ever experience! They had never seen death, had never even imagined it. It must have been for them like suddenly finding themselves in the starring role of the most gruesome and grotesque horror show you or I have ever seen!

If the purpose of their watching the slaughter of the innocents was education, the Lord must have wanted the humans to know how to take the life of a fellow creature properly, which one(s) to choose, how to turn bloody skins into supple hides, and all such as that. There would have been several possible reasons for this lesson in slaughtering and butchering: one, going forward, the humans would have no choice in some cases but to kill their fellow creatures in order to get enough food to survive in the wild. Besides, the humans, once outside the protection of the garden's ecosystem, might become food for the animals, too. So it was a kill-or-be-killed situation into which they were about to be thrust. Whether this curse went into effect immediately or came later is debatable. We will return to that question later. Two, because of sin, God would now require an atoning sacrifice. Although the narrator never says that God would begin requiring a sacrifice at that moment, we can and should infer it from what follows at the beginning of Chapter 4, as Cain and Abel bring offerings to the Lord. Consider that sin had brought the other consequences already mentioned rather immediately: for the woman, sex caused ovulation and menstruation, followed by pregnancy, morning sickness, pain in childbirth, and constant maternal care to keep a newborn alive; for the man it had caused the never-ending struggles of providing food, clothes, and shelter for his family by the sweat of his brow, and of defending his family through whatever perils life in the wild brought. Would not sin, therefore, create the immediate need for a sacrifice of atonement? It is evident from everything we see in the Bible—from the Cain and Abel story of Genesis Chapter 4 all the way into the New Testament, that God required some kind of sacrifice, so surely this requirement began at the time of the Lord's cursing the whole earth when addressing Adam. Indeed, the animal kingdom as a whole, not just the snakes, would be forever cursed because of the sin of the humans!

This raises a question, though. Why, if God required a sacrifice of atonement for sin did it have to be the murder of a helpless animal? Could it not have been something less gruesome, such as making the man bow and pray five times a day, fast every Thursday, dig holes only to fill them back in, write on a cave wall a hundred times "I will not disobey the Lord again," or some such act of penance? Wouldn't any such punishment do the trick? It seems theoretically that God could have used anything. Yet He chose an animal sacrifice. The explanation for that choice is this: because the Lord himself killed a helpless creature just to make clothes for Adam and Eve, He wanted them and their descendants to have to do the same in perpetuity as a reminder. The sacrifice, therefore, was not meant merely as a punishment; it was also meant to be a memorial. God works that way. We know that from later Bible stories, such as how He required the Israelites to keep the Passover, and how Jesus told his disciples to practice communion in remembrance of him. It also served to show the humans that their actions not only affected themselves but all of creation. To the first two humans, this memorial act had the intended impact. They would never be able to forget its purpose. They would feel guilt, shame, and remorse every time they performed a sacrifice, and they would be reminded of the consequences of disobeying the Lord. One would think that Adam and Eve would have passed down the meaning and purpose of sacrifice to their children, grandchildren, great grandchildren, and so on, so that they, too, would keep the memorial from generation to generation. That may not have been the case, however, as we will see in the next chapter. At some point over a long period of time, it is obvious that the purpose of sacrifice would be lost/forgotten, and in time the fallen angels would begin to interact with the humans and demand sacrifices, but they would do so for the sinister purpose of wanton destruction of God's creation (animals) and for the deception and corruption of the humans. Moreover, in time, these "gods" would even convince the humans to murder their own kind as sacrifices. And Satan would laugh, still naively hoping somehow to outsmart God and thus make the species created in his image and likeness go extinct.

To conclude our discussion of Chapter 3, we see the Lord speaking in the plural again, just as He did in Chapter 1 verse 26, saying the humans have now become like "us," knowing the difference between good and evil. The same question arises here as arose before: who is "us?" In keeping with the explanation proffered earlier, we must assume He was talking primarily to the non-earthly beings that believers call angels and which many non-believers call extraterrestrials. He (the Lord) could have been talking to his Father (the Elohim) as well, but since the Elohim is spirit rather than flesh-and-blood, it is more difficult to visualize that than to suppose He was talking to the angels. Either way, He stated matter-of-factly that the humans' innocence was gone, and it concerned him that they might now take of the tree of life just as they had disobediently taken from the tree of the knowledge of good and evil, thereby living forever in their corrupted state. Therefore, He took two actions to prevent that. One, He put the humans outside the Garden of Eden, which implies that the garden was an enclosed space of some kind. Whether enclosed with walls or fences, or enclosed with some kind of natural barrier such as mountains, rivers, or oceans, or perhaps even enclosed within a bio dome, we cannot know. Two, He put guards at the entrance to the garden to make sure that the humans didn't come back in.

Since every single word in the story seems important when making a thorough analysis of Genesis, we must assume that it is important that the narrator tells us the entrance lay on the eastern side of the garden. He did not say that the humans were put out on the eastern side, however. They may have been put out on the side farthest from the entrance. Regardless, we must assume it to have been possible for the humans to make their way back to the entrance in time; otherwise why guard it? The Lord chose to guard it with other-worldly beings called cherubim. The narrator does not explain why cherubim as opposed to some other kind of angelic beings were chosen for the job. Nor does He tell us what made/makes cherubim a different order of beings than other angels. Meanwhile, the Lord instructed them to guard the tree of life, or at least the "way *of*" it, by which it is generally assumed that the narrator means the "way *to*" it. Whether the narrator is referring to a road, a bridge, a pathway, a hallway, or something else, we cannot know. Either way, the Lord chose to have it guarded with a "flaming sword which turned every way." This sounds like some kind of high-tech defense mechanism. It is difficult to visualize anything that currently exists as being representative of this. We can use our imagination and picture fences of lasers or of electricity, perhaps, but nothing quite suffices. We must be content to know that in the celestial realm, technologies exist that humans have not yet conceived of and, it may be, that we *cannot* conceive of.

The actual meaning of the words "way of" may not be "way to," anyway. It seems more likely that the narrator was trying to say that the Lord would permanently block the access of humans to the ability/opportunity to create life by non-sexual means; i.e., by the scientific laboratory means called the "tree of life." Now that Adam had chosen sex over obedience, the Lord decided to make that the means for reproduction of the species from then on. Adam and his posterity would forever be stuck with it as the natural consequence of that choice. Over time, later generations completely lost the knowledge that there could have ever been creation by any other means than sex. Why later generations lost that particular piece of knowledge, not to mention countless other pieces, will become clear in the next chapter.

Meanwhile, it should be noted that when the Lord put Adam and Eve outside the garden, He cut them off from all technology. He thrust them into a caveman-like existence, which is in fact precisely why we see cavemen in our archaeological past. The first man and woman were completely destitute, having no weapons, no torches, no shoes, no anything, we must assume, except the skins they were wearing. Perhaps the Lord helped them temporarily with some barebones essentials just to ensure their survival, but it is clear that He expected them to do for themselves as soon as possible and to the greatest extent possible. They had none of the amenities of life that we enjoy today. Life for them would be hard. To us, it would seem unbearable, so spoiled are we by modern conveniences. They would from then on live in a kill-or-be-killed by wild animals state of existence. Since there was no way for them to spin thread or weave cloth, the butchery of animals would provide their clothes. And since there was no way for them to forge metal tools for hunting, slaughtering animals, plowing, planting, or reaping, they would have to find and use sharp stones, pointed sticks, and whatever else they could figure out to use. Over time, they or their descendents would develop metallurgy, but for many years—probably decades if not centuries—these first humans lived in the Stone Age.

Genesis: The Explanation

What Really Happened
"In the Beginning"

4

THE EXPLANATION
GENESIS Chapter IV

SURVIVING THE STONE AGE

1 Now Adam knew Eve his wife, and she conceived and bore Cain, and said, "I have acquired a man from the LORD." 2 Then she bore again, this time his brother Abel. Now Abel was a keeper of sheep, but Cain was a tiller of the ground. 3 And in the process of time it came to pass that Cain brought an offering of the fruit of the ground to the LORD. 4 Abel also brought of the firstborn of his flock and of their fat. And the LORD respected Abel and his offering, 5 but He did not respect Cain and his offering. And Cain was very angry, and his countenance fell. 6 So the LORD said to Cain, "Why are you angry? And why has your countenance fallen? 7 If you do well, will you not be accepted? And if you do not do well, sin lies at the door. And its desire is for you, but you should rule over it."

8 Now Cain talked with Abel his brother; and it came to pass, when they were in the field, that Cain rose up against Abel his brother and killed him. 9 Then the LORD said to Cain, "Where is Abel your brother?" He said, "I do not know. Am I my brother's keeper?" 10 And He said, "What have you done? The voice of your brother's blood cries out to Me from the ground. 11 So now you are cursed from the earth, which has opened its mouth to receive your brother's blood from your hand. 12 When you till the ground, it shall no longer yield its strength to you. A fugitive and a vagabond you shall be on the earth." 13 And Cain said to the LORD, "My punishment is greater than I can bear! 14 Surely You have driven me out this day from the face of the ground; I shall be hidden from Your face; I shall be a fugitive and a vagabond on the earth, and it will happen that anyone who finds me will kill me." 15 And the LORD said to him, "Therefore, whoever kills Cain, vengeance shall be taken on him sevenfold." And the LORD set a mark on Cain, lest anyone finding him should kill him. 16 Then Cain went out from the presence of the LORD and dwelt in the land of Nod on the east of Eden. 17 And Cain knew his wife, and she conceived and bore Enoch. And he built a city, and called the name of the city after the name of his son—Enoch.

18 To Enoch was born Irad; and Irad begot Mehujael, and Mehujael begot Methushael, and Methushael begot Lamech. 19 Then Lamech took for himself two wives: the name of one was Adah, and the name of the second was Zillah. 20 And Adah bore Jabal. He was the father of those who dwell in tents and have livestock. 21 His brother's name was Jubal. He was the father of all those who play the harp and flute. 22 And as for Zillah, she also bore Tubal-Cain, an instructor of every craftsman in bronze and iron. And the sister of Tubal-Cain was Naamah. 23 Then

> Lamech said to his wives: "Adah and Zillah, hear my voice; wives of Lamech, listen to my speech! For I have killed a man for wounding me, even a young man for hurting me. 24 If Cain shall be avenged sevenfold, then Lamech seventy-sevenfold." 25 And Adam knew his wife again, and she bore a son and named him Seth, "For God has appointed another seed for me instead of Abel, whom Cain killed." 26 And as for Seth, to him also a son was born; and he named him Enosh. Then men began to call on the name of the LORD. [NKJV]

The narrator begins Chapter 4 by telling us in verse 1 that Adam indulged in sex with his wife Eve. He did not say "again." Therefore, he does not tell us whether this particular act of sex was the first time, second time, or one of dozens (or hundreds? even thousands?) of times thereafter. It is possible, therefore, that this particular act was the first time, and the narrator is referring back to the original sin in the garden in Chapter 3. If so, Eve became pregnant as a result of that original sin. We may be tempted to suppose, however, that the narrator is implying that after the original sin in the garden, Adam and Eve had sex repeatedly as part of their new daily lifestyle. But that may not be true, for a reason that will become evident shortly. Either way, Eve got what she wanted—pregnant—and she soon felt the pain that the Lord informed her would come as a *consequence* of natural childbirth (not *punishment* for committing sin—meaning having sex). The narrator tells us nothing about what happened during the nine months that the first human fetus began growing inside the first mother-to-be, but we should assume that a few things occurred. One, Eve probably vacillated between regretful shame, eager anticipation, and fear of the unknown. Two, her maternal instinct surely kicked in, and she somehow just knew what to do to prepare for having a child, just as the animals always did.

Three, Adam likely was more filled with wonderment and worry than anything else. What would a human baby look like? What would it act like? What would be his fatherly responsibility toward it? Certainly, he was concerned with how he would provide for the child; after all, life in the wild for himself and Eve was hard enough, everyday being a struggle for survival. How would he defend a vulnerable child against predators, when he and Eve had to look over their shoulder at all times just to protect themselves? What preparations should he make for living with another human? Beyond that, Adam seems to have played a mostly passive role in this drama of pregnancy and childbirth; he was basically just an onlooker. Protector and defender, yes. Supportive husband, certainly. But action figure in this drama? Hardly. The fact that he is in the background rather than acting as the main character is full of symbolic significance. It bolsters the case made in the previous chapter for Eve's being the real catalyst for change, the one who by forethought pushed for having sex and making babies like animals do, regardless of what the Lord had told Adam. Now she would take on most of the responsibility for parenthood and family structure, while her husband would be thrust into the family life reluctantly. Ultimately, regardless of how either parent-to-be felt about the situation, they were now irreversibly stuck with it and would have to make the best of it.

FAMILY LIFE

When the time came, Eve went into labor and suddenly knew what the Lord meant when He said human childbirth would be painful. In between contractions, she undoubtedly felt remorse for her disobedience. And with each agonized push, scream, and tearing of her flesh, Adam broke down and wept, pleading with the Lord to have mercy. Did the Lord officiate at this birth, like a supervising physician? Did He send angels as midwives? Or did the Lord imbue Adam with the instinctive know-how to be in charge? We cannot know. Perhaps Adam watched dumbfounded while Eve did it all alone. Or perhaps he couldn't watch and had to turn and rush away in angst. What we can know is that eventually, Eve pushed the large head and shoulders of her baby through her birth canal and fulfilled the desire that had brought about the original sin; she successfully gave birth to a live human!

The narrator tells us that Eve's pronouncement upon giving birth (presumably upon holding her baby in her arms for the first time) was something along the lines of, "I have gotten a man from the Lord." But it would be foolish not to wonder whether she also said something along the lines of, "Tree of life? Who needs a tree of life? *I am* a tree of life!" She may not have voiced that thought openly, and it is not even implied in the narrative, but in light of all the drama that came before it, it stands to reason that this thought probably crossed her mind. At that moment, far from regretting her decision to make a baby the animalistic way, inside she surely felt triumphant. Adam undoubtedly recognized this maternal pride in his wife, and he probably reacted to it with bewilderment, but if so, he had little time to take it all in and decide how to feel about it. He was immediately seized with awe as he beheld the first human besides himself and his wife. If his reaction upon meeting Eve was, "This is bone of my bones and flesh of my flesh," how much more so must his reaction have been to meeting his first-begotten son! Still today, when fathers see and hold their newborn for the first time, we stand there in awe, head swimming, heart swelling, emotions swirling; we are rendered speechless. Imagine how Adam must have felt all of that but to a much greater degree.

It is implied that Eve, not Adam, named the newborn. That makes sense, because clearly, this was the product of her active choice; it was her love child. It had been her idea to have a baby the same way the animals did, and now she had one. She named him "Cain" (or "Qayin" in Hebrew), meaning something like "what I got," "what I received," or better yet perhaps "gift." That this gift came as a result of the Lord's mercy seems both obvious and incidental, not really all that significant in the story, yet the narrator tells us that Eve made a point of acknowledging it. Why? Was she really wanting to honor God at this moment? Or did she perhaps actually want to cloak her true feeling, which, as mentioned above, was one of "Look what *I* did; *I* created life!" We cannot know for sure, of course, but it would be reasonable to assume the latter, because it seems within our human nature—which we all inherited from mother Eve and father Adam—to feel pride inwardly for our accomplishments, even if outwardly we express humility. So, perhaps it was false humility, or at least the tug-o-war between pride and humbleness that goes on within us all that caused Eve to acknowledge the Lord here. Either way, to some extent, cloaking her pride was tantamount to covering herself with another fig leaf. If Adam agreed with Eve's outward statement, as we should presume he did, that the baby was a gift from the Lord, while simultaneously recognizing in his wife an unholy maternal pride, and yet he remained silent, then he was covering the nakedness of his own heart with another fig leaf as well. Again, it must be stressed that the Bible does not say any of this, so it is pure conjecture, but it helps fill out an otherwise tacit and terse narrative within the overall story of Genesis and make it line up with observation of human nature today.

After the baby was born and named, the minutes passed into hours, and the reality and responsibilities of life soon began to replace the joy and exultation of having created life. The fear of the unknown returned with a vengeance, and the harshness of life in the wild brought Eve and Adam back to earth. Daily activities of hunting, gathering, planting, reaping, and perhaps fishing took precedence over all else. This humdrum of life continued day after day, year after year, but it was punctuated by times of sexual enjoyment. How frequently we cannot know. Again, our temptation in modern times is to assume that, lacking any other form of entertainment in the primitive state of life outside of Eden, there was little else for them to do for pleasure. Yet considering the shame they had experienced from the act of losing their virginity, and knowing the Lord had not been pleased with them, they may have tried to suppress the urge for long periods of time before caving in to the desire, only to find themselves plagued by guilt thereafter. And this cycle may have repeated numerous times, just as it still does for many people today. Regardless, soon enough, Eve became pregnant again and gave birth to another son, which she or Adam named Abel. Whether Eve or Adam named this child is irrelevant to the story, but we may do well to suppose that Adam named him, because except for Cain, Adam had named all other living creatures he had encountered; it was his divine prerogative. There is another reason to think so, too, which will become evident shortly.

"Abel" in Hebrew means something like "vapor," "breath," or "spirit," and some scholars speculate that this word may actually be a description of this character's short life rather than his given name. We might be wise to take it as his given name, though, and understand it to mean "a living, breathing being." Whether Abel was in fact the second child born to Adam and Eve is debatable. Some scholars have speculated that Cain and Abel actually might have been twins. While that way of interpreting the scriptures is not necessarily outlandish, it does not seem likely. What does seem likely, or at least possible, is that one or more girls might have been born between the births of Cain and Abel. It is common knowledge that the value placed on girls in society all through ancient history was low. It was low enough that the narrator scarcely mentions any by name after Eve until we get to Sarai in Genesis Chapter 11. Yet we know from all available evidence that the proportion of female to male births has been close to 50-50 all through history, and we know from Genesis 5:4 that Adam and Eve had some girls. What we cannot know is how many, but we can assume they had about the same number of girls as boys. Hebrew tradition says that they had 33 sons and 23 daughters in all over a span of many hundreds of years. There is no way to know, of course, but just because the Bible mentions only their first three sons by name does not mean they had no others. Considering the long life spans of the first humans, the 33 and 23 numbers seem plausible.

THE FIRST DEATH

Where was the Lord all this time? The Bible does not say, but we know He was close by, still on earth or at least visiting frequently in his physical form. We know that because of what comes next in the story: the Lord was physically present to receive or refuse offerings made by Adam's sons, and to talk to Cain about it. The story starts with the fact that, over the course of time, Cain became a farmer, and Abel became a shepherd. One day, each presented the Lord with a gift derived from his labor. Thus, Cain gave the Lord some produce, while Abel gave the Lord some stock from his herds. Although we typically picture these offerings as sacrifices made ceremonially upon an altar, the Bible does not say that is what happened. We may just as well picture that the Lord—a physical being walking the earth while visiting it—had to eat, just as Jesus did thousands of years after the events recorded in Genesis, so Cain and Abel literally cooked for him. They prepared for him the same foods they themselves ate, but they of course focused their respective menus on the items they had worked hard to raise. Perhaps between the two of them, they expected to present the Lord with a well-balanced meal! And perhaps they worked together, not separately, to make one presentation in two parts. Each man was hoping for and expecting to receive validation from the Lord—pats on the back, if you will—for a job well done. But instead, the Lord gives Abel an "attaboy" and gives Cain the cold shoulder.

This story has provided about as much fodder for unbelievers, skeptics, and apologists alike to argue over as anything else in the Bible. To unbelievers, it seems evidence of a mythical, capricious God's favoritism toward one guy and discrimination against the other. To skeptics, it seems to point to the Bible's being confusing and contradictory, if not downright baffling and incredible. Both Jewish and Christian apologists meanwhile frequently tend to tie themselves in knots trying to explain away this mysterious story. So what is the actual explanation? Let us first establish what it is not. The explanation is not the one proffered so commonly in commentaries—that Abel gave the best of his flock as a sacrifice while Cain gave something less than his best. Nor is it the idea so popular among unbelieving commentators that the story of Cain and Abel represents the tension between the stationary farmers and the nomadic herders of antiquity. Likewise, it is not the odd belief that God has an insatiable appetite for blood, or its corollary, that He is a carnivore and just doesn't like vegetables! Finally, it is not the alternative theory that Cain wasn't the son of Adam; he was the son of the serpent (and by extension the son of Satan). No, the explanation can be found in the preceding chapter in our discussion on the Lord's killing of animals to make skins with which to clothe the naked and ashamed humans.

Thereafter, the Lord instituted sacrifice to serve as a reminder that human disobedience to God brings death to all living creatures, not just to humans themselves, and to show the humans how to humanely kill and prepare meat for the table in their life outside the garden. It also served to help the humans learn to cope with death, because they would now be seeing a lot of it.

If that is the case, then the Lord rejected Cain's offering because it was the wrong *kind* of offering, not because it was an acceptable kind which was merely carried out poorly due to Cain's failing to give the choicest produce from his harvest. Think of it this way: in presenting the wrong kind of offering, Cain was basically trying to give the Lord something the Lord had not instructed humans to give him. In that sense, Cain was approaching sacrifice as a religious ritual rather than as an act of obedience to the Lord's instructions. This is, not insignificantly, the root of all religious rituals throughout history. We humans want to "do" something to win God's favor. We want to impress him by our efforts. In Cain's day, that meant making a ritual sacrifice while missing the prescribed *purpose* of sacrifices. Later in history, by the time of Christ's ministry, it meant Jews keeping the letter of the law while missing the point of the law. Still later, in the age of Roman Church dominance, it came to mean elevating a man to the status of an infallible Pope and elevating an institution beyond criticism while silencing the critics with threats of torture or death. Since the Protestant Reformation, it has come to mean straining doctrinal gnats while swallowing camels of religious pride and sowing discord in the body of Christ to try to prove we are "right" and our fellow Christians are wrong. Today in America, it means all too often that we put on a religious performance for the Lord in church every Sunday only to go home and slouch back into our comfortable lives of fleshly excess for the rest of the week.

Getting back to why the Lord rejected Cain's offering, remember that He didn't demand sacrifices because He was hungry, even if He did normally accept and eat the food offered to him by the humans. Nor did He demand them primarily because He wanted to see if humans would voluntarily deprive themselves of the fruit of their own labor by sharing it or giving it away, although that certainly sounds very proper to our Christian sensibilities today and may very well have been a secondary reason. We know it was not the primary reason, though; if it had been, the Lord presumably would have accepted Cain's offering, if only reluctantly, with a qualifier like "this is okay, but you should try to do better next time." We know the Lord rejected it outright, so we must therefore conclude it simply was the wrong kind of offering, and as such the Lord couldn't accept it at all—not *wouldn't* accept it, but *couldn't*.

It seems that Cain either misunderstood the purpose of sacrifice, or else he understood it perfectly well but took it upon himself to make a change in how sacrifices were conducted without consulting the Lord about it. Let us suppose the former was the case. Why would he have misunderstood? Consider the following as a possible answer. The Bible does not give an account of any sacrifices being made prior to Cain and Abel's in Genesis Chapter 4, but Adam and Eve surely made sacrifices, and not a few. Let's assume Adam and Eve made a sacrifice only once per year, Cain was born in the first year, and children matured at the same rate back then that they still do today. They would have performed some fifteen, twenty, or more sacrifices before Cain was old enough to begin asking the reason for making them. If Adam and Eve made sacrifices twice per year—say on the solstices (just for argument's sake)—then we are looking at thirty, forty, or more which Cain watched them make. He probably even helped them with the process once he got old enough, and increasingly so from year-to-year thereafter. Yet, despite all of this, if the original sin that brought about the need for animal sacrifice was human sexual relations, father Adam and mother Eve may not have told Cain (or any of their other children, for that matter) the real reason for performing them. Why? Because if Adam and Eve were ashamed of their nakedness in the presence of the Lord after their initial sin, why would they have been any more comfortable discussing nakedness and sex with their children? We might do well to assume that they told Cain that their disobedience to the Lord caused the need for sacrifice without specifying what particular act of disobedience started it all. In other words, perhaps they never actually explained to Cain the connection between their sexual sin which caused the Lord to kill an animal to make clothes for them and their performance of blood offerings. To Cain, therefore, the fact that they made clothes from the hides to cover their nakedness was incidental, not the main reason for killing the animals in the first place.

If so, that brings up another question: if the Lord himself was still walking among this first family of humans on earth from time-to-time, why did He not tell Cain the reason himself? The answer is because He operated the same way in this case as He did when He placed Adam in the Garden of Eden. Remember that He gave the prohibitions against partaking of the trees to Adam, not to Eve. It was Adam's responsibility to instruct Eve. In this case, it was Adam and Eve's responsibility as parents to instruct their children, for the Lord clearly set up the timeless, universal principle of parental responsibility from the start. (And as generations passed and the human population grew, the Lord expected societal responsibility for proper moral instruction to supplement parental instruction.) Moreover, if they didn't tell Cain, then we must assume they didn't tell any of their other children, either. And if that is the case, no human ever born on earth after them could have ever known it since the Lord did not intervene to tell them. Even the narrator of Genesis and later authors of Biblical books wouldn't have known it. Who probably did know it, however, was Satan and the fallen angels who were banished to the earth. They gladly used Adam and Eve's parental ineptitude against them and their progeny, over time convincing humans that sacrifice should be made to them as their "gods"—those who visited them from the sky from time-to-time. Thus was the mysterious phenomenon of animal, and even human, sacrifice spread throughout the antediluvian world. Even after the flood, the practice of animal sacrifice was restarted by Noah and his descendents, with the same ignorance of original purpose. The lack of understanding of the true purpose of sacrifice was thus hidden for millennia, all the way up to the time of Christ's death on the cross, whereupon it ceased to be necessary to perform animal sacrifices. Over the centuries since then, as most nations have one-by-one stopped practicing animal sacrifice, it ceased to be necessary to inquire into the original purpose.

If this piece of the puzzle is fitted properly, then Adam and Eve simply failed . . . again. Whether they were altogether derelict in their duty or they simply didn't do a good job of instructing their children, they certainly didn't live up to the Lord's expectations in that regard. Why should we suppose that after their initial sin, they suddenly did the right thing in every other case for the rest of their lives? Would it not be more prudent to assume they made parenting mistakes just like all the rest of us? And would it not be wise to bet they messed up in other ways, too? The bottom line is, they may have left Cain without proper education about how to perform sacrifices, and thus in his first attempt at an offering in which his parents were not participants, he did it incorrectly.

Now let us consider Abel. The Bible does not tell us when Abel was born or how much younger he was than Cain. It seems possible, if not likely, that Abel was only about ten months to one year younger because of Adam and Eve's divine genetic composition. Adam surely would have been as virile a male and Eve as fecund a female as any humans have ever been. With their newlywed passion in full bloom, they might have produced a small family rather immediately before they learned to curtail their sexual activity. (We know that they did indeed learn to curtail it soon after Abel because Genesis 5:3 tells us that Adam was 130 years old when his next son, or what we must presume to have been his next son, Seth, was born.) If that is the case, then their second child probably arrived within roughly a year after the first one. Whether that child was Abel or an unnamed sister, we cannot know. Let us assume, however, it was Abel. If Abel was that close to Cain in age, he would have been Cain's equal in many things, his near-equal in some things, and probably his superior in others. Sibling rivalry was surely a major dynamic in their relationship from the time they were toddlers. If so, Abel would have been privy to the same information as Cain about offerings and the performance of sacrifices. Therefore, we can assume that he didn't know the real reason why animals had to be killed, either, and the fact that he offered an animal on the occasion discussed in Chapter 4 had nothing to do with his being more obedient or knowledgeable than Cain. He simply offered an animal for the same reason that Cain offered produce—it represented his career of choice, the work of his hands, the fruit of his labor, the result of the sweat of his brow.

Moreover, it is possible that, even though the story in Genesis makes it seem as though Cain and Abel perform their sacrifices simultaneously, that may not have been the case. It may be that, despite Abel's being younger than Cain, he actually made his sacrifice first and Cain was playing catch-up to little brother. In this scenario, Abel made his offering, the Lord received it, then Cain, out of jealousy, rushed to put an offering together. It may not have occurred to him that he needed an animal, when, after all, he had all that wonderful grain! If it did occur to him, surely his pride did not allow him to ask Abel for a lamb or goat. So, according to this scenario, he put together a hastily-prepared sacrifice out of jealousy which contained the wrong kind of offering. Whether true or not, regardless of whether pride and/or jealousy were involved, he certainly made the wrong *kind* of offering.

Moving on, let us now parse the next scene in Chapter 4: the Lord's scolding of Cain and Cain's response to it. The narrator does not tell us the specific chain of events in the order they occurred, but in verse 5 he does tell us that Cain got angry as a result of the Lord's accepting Abel's offering and rejecting his. The implication is that Cain was present or within eyesight as Abel and the Lord enjoyed a pleasant exchange. However, we could speculate that Cain was not there and merely heard about it after the fact. We can envision a pretend dialogue between Cain and Abel after that offering day that went something like this:

"Hey little brother, the Lord didn't accept my offering, and I don't know why. Can you believe that?"

"Wow, that's too bad; he accepted mine. Said it was just fine."

"What the . . . ?"

"Well, I guess mine was better than yours. Maybe if you do a better job next time . . .".

Even though it is possible to imagine that kind of dialogue, it probably didn't occur *before* Cain got angry but *after* he was already upset from watching the Lord accept Abel's offering. Indeed, the way the story probably played out puts this type of brotherly conversation in verse 8, because the events of verses 3 through 7 seem to have all occurred one after the other on the same day. If so, the chronology goes like this: Cain and Abel present their sacrifices back-to-back on the same day, the Lord approaches Cain's first (since he was the firstborn) and rejects it, then goes immediately to Abel's offering and accepts it; Cain sees the Lord pat Abel on the back, is dismayed, gets mad and feels jealous; the Lord senses that Cain is upset, admonishes him for it, and warns him that he is treading on thin ice, and does all this right in front of Abel. The Lord then walks away, leaving Cain standing there dumbfounded and Abel feeling like a million bucks.

Soon after, probably on the same day, Cain and Abel have the brotherly conversation mentioned above, but it actually goes like this: "Dang, little brother, I just don't understand. I can't believe the Lord accepted your offering and rejected mine. He said if I 'do well' He would accept my offering next time. What the heck is He talking about? What was wrong with this offering? He didn't say, and I was scared to ask. He seems so hard to please..." Cain's shoulders drop in despair, his eyes fall to the ground, and he shakes his head in bewilderment, "... But He accepted your offering! What did He say to you?"

"He didn't say anything special, just that He was pleased with my work. That was good enough for me. Man, I worked my tail off getting that offering ready for him. I don't know what was wrong with your offering..." Abel shrugs, makes a nonchalant hand gesture, raises his chin, looks down on Cain, and begins turning away... "I guess mine was just better than yours. You know how you get careless sometimes. Maybe you should show more attention to detail next time."

Cain immediately reacted to Abel's dismissive arrogance: "How dare you! You little...! Who do you think you are? What makes you think you're so special?" [his former dejection switching to anger. Then just as quickly he remembered, "Oh, the Lord, that's what," and his anger devolved into wounded pride and jealousy.] As he stewed on all this, over a matter of days he and Abel had more conversations as part of their daily routine, and his anger grew in proportion to Abel's haughty demeanor. He eventually got to the point of deciding he had to teach his little brother a lesson, although he probably didn't intend to kill him.

That brings us now to the second half of verse 8. At some unspecified point after talking to Abel in the field, Cain killed him. How, we cannot know. Whether it was premeditated murder or an accidental homicide is also debatable. It is easy to visualize the two young men having a heated exchange then getting into a tussle, but beyond that it's anyone's guess. For the sake of moving the story along, let us just assume that Cain threw Abel to the ground and bashed his head against a stone. Let us also assume that Cain did not intend to kill Abel but just to beat him in a hand-to-hand bout in order to regain some of his dignity and to serve his little brother a slice of humble pie. Keep in mind that up to that point in time, no human had ever died, no human had ever witnessed another's death, and all humans must have even wondered what the death of one of their own would look like. Adam and Eve certainly would have wondered about it, because the Lord had told Adam specifically that sin would result in death, and he had passed that information along to his wife. Whether the couple had told their children about death is yet another unanswerable question. It may be that Cain didn't know it was possible that one human could take the life of another, although it is hard to imagine that, because he saw the killing of animals on a regular basis.

Pausing for a momentary digression and backtracking to verse 2, let us consider a point not previously discussed about Cain and Abel's chosen professions. It is possible that Cain chose to be a "tiller of the ground" because he didn't have much of a stomach for killing animals. Perhaps he was even the first of a type of the human race that today we know as vegetarians, or of the type we know as animal rights advocates. Perhaps it bothered his conscience to see all of the killing, even though he partook of the meat and wore the skins. If so, it seems a supreme irony that it would be he who took the life of the first human; he wouldn't kill an animal if he could avoid it, but he would kill a human if necessary. This, like so much else in this study, is pure conjecture, but since we cannot know Cain's motivations, it is as worthy of our consideration as anything else. (Interestingly, that same dynamic is sometimes present today in the circles of radical animal rights activists and environmental conservationists; some would rather that their fellow humans suffer and die than that an endangered species go extinct or a wetland get disturbed!)

EAST OF EDEN

The tragic killing of Abel in verse 8 gives way abruptly to the Lord's questioning Cain about it in verse 9. As was the case when the Lord questioned Adam and Eve about their sin in Chapter 3, we can only speculate that the Lord did not ask Cain, "Where is your brother?" because He genuinely did not know, but rather because He wanted to give Cain the opportunity to admit his sin. And as happened with his parents in the garden, he did not admit it at first. He pretended not to know, as if the Lord wouldn't call him on it. But the Lord did call him on it, of course. When Cain answered, saying in essence, "I don't know where he is; I'm not his babysitter!", the Lord immediately retorted, "What have you done? The voice of your brother's blood cries out to me from the ground." The Lord then proceeded to place a curse on Cain, telling him that the very dirt beneath his feet, which had just soaked up his brother's blood, would now be his enemy. It would not yield its produce to him easily anymore. What a terrible punishment it must have been for a young man who prided himself on his farming and gardening ability! Remember, he chose to sacrifice the fruit of the field to the Lord rather than kill an animal, so the one thing in life that was nearest and dearest to his heart was the produce that came from putting his hands in the dirt. Yet, as bad as that was, the Lord had even more punishment in store for him. Before Cain could even mentally process the first part of the curse—as he stood there astonished, with his eyes wide and his jaw agape—the Lord told him that he would also be kicked out of his familiar surroundings and sent wandering into unknown lands.

This punishment was very much the same thing the Lord had told Cain's parents when He expelled them from the garden: if you sin, I will remove the blessing of a safe, comfortable home environment from your life; I will make you homeless and destitute, and you will have to start your life all over again. The message the Lord was sending to them in both cases was, "I made you from the dirt of the ground under your feet, and I then gave you the ground under your feet as a solid foothold in life—a place for you to live in safety and comfort. And yet you disobeyed me, which means you think you know better than I do what is okay and not okay. Let me show you how little you know. I will shake the very ground under your feet, and send you to find a new homeland in the wild, untamed wilderness. Then you will see that I have been protecting you, only to have you take that protection for granted. I will remove my protection from your life and let you fend for yourself. That will get your attention. That will make you realize that you can't just do your own thing. That will cause you to realize that you cannot live by your own thoughts and efforts, but only by my word." To Adam and Eve, He said, "You thought my rule about not partaking of that one forbidden fruit was unreasonable, but I will show you how good you had it." To Cain, He said, "You thought my rule about how to perform sacrifice was onerous. I will show you how easy it would have been to enjoy your life."

Cain's reaction to his punishment in verses 13-14 was one of being confounded, if not altogether astonished. He could not believe his Lord would be that hard on him. When we read this story today, however, we cannot conceive of why Cain would think that; it seems obvious to us. If one person murders another person, the murderer must be punished if there is to be any justice in the world at all. Anything other than a severe form of punishment, whether capital or something else, would be a grave injustice. If that makes so much sense to us today, why did Cain not see it that way back then? Or did he actually see it but feign his reaction, hoping to obtain some kind of mercy from the Lord? Either could be the case, so let's explore each one briefly. If he was indeed incredulous about his punishment, it must have been because no one had ever killed anyone before, he had nothing to compare this type of sin to, and thus he had no standard in his mind of what *ought* to happen to a murderer. In addition, he may have even rationalized that, "Well, humans kill animals all the time; animals kill other animals all the time. Why should I be punished for what I did if, say, a lion is not punished for killing a deer?" Although that may sound silly to us today, it may very well have been the case with Cain. Besides, we should all be able to relate to the rationalization of our sins. So let's not think we are so very different from Cain.

Now let's suppose Cain was feigning incredulity. Would *that* make him much different from us today? After all, any of us can be quite manipulative when we find ourselves stuck in a situation as uncomfortable as his. Most of us would throw ourselves on the mercy of the court, beg for a lighter sentence, and try to sway the judge by playing on his/her emotions. Is that not what Cain was doing here? That much seems clear. Whether this manipulation was truly meant to be deliberately deceptive, however, is less clear. Cain may have been guilty of something else which any of us can fall victim to when put in a bind like this: self-deception. Although he may have known deep down in his heart that he wasn't being completely honest with the Lord, he may have shut down that part of his conscience in order to plead for leniency. The fact is, all of us are capable by our human nature of doing the same thing in similar situations. Our instinct is for self-preservation above all else, including above being truthful with ourselves. As a reflex, we mentally suppress what we know to be true in order to achieve an outcome that will lead to self-preservation. Over time, as self-preservation has been achieved, and we feel the danger has passed, then we begin to dig through the layers of our minds to unpack the self-deception. The point is, whether Cain was feigning incredulity or not, his reaction was not all that different from how most of us would have reacted in a similar situation.

Cain immediately laid out a list of reasons why his punishment is too severe: 1) it is greater than he can bear, 2) he will be hidden from the Lord's presence, 3) he will be homeless and unwanted by anyone, and 4) other people will want to kill him. Let's look at each of these reasons now. The first thing on the list seems to encompass all the others. However, it could be that Cain was really displaying the first case of depression in history. In other words, he may have been saying that his spirit would be so crushed that he could not carry on, which is another way of saying he might end up committing suicide. The second thing on the list is a complaint that he will be sent away from the presence, fellowship, and protection of the Lord and/or other humans, such as his parents. Third on the list is his fear that he will never be able to rest, relax, or enjoy his life, because he will be alone and unwelcomed by anyone. Fourth is his fear that someone will try to hunt him down and kill him, undoubtedly as revenge for the murder of Abel. But if so, who are these other humans? There was no one yet alive that we know of except Adam, Eve, and perhaps a sister or two. Surely he didn't fear them, did he? Was he projecting into the future, thinking of his brothers yet to be born? Or was it the angels or the fallen angels he feared? We will return to this topic shortly.

    The Lord's response in verse 15 to this litany of complaints is to ignore numbers one through three and to address only number four. In ignoring the first three, the Lord was essentially saying to Cain, "Suck it up. Get over it. You'd better be glad I don't execute you myself, or send you to be executed by your father, Adam." In addressing the final one, it seems clear that the Lord wanted to preserve Cain's life, and in fact was not going to let anyone take revenge on this first begotten human. So the protection that Cain feared he would lose turned out to be unfounded. The way the Lord chose to provide that protection has, of course, been a subject of great mystery ever since. He put a "mark" of some kind on Cain. Speculation about what that mark was could fill a separate chapter, and it could also distract us from more important aspects of the story. Consequently, we should look it at only briefly. At one time it was fashionable to think that the mark of Cain might have been that the Lord changed his skin color from white to black. Without any equivocation, that antiquated and absurd idea must be rejected outright. First of all, there is nothing in the Bible to suggest that when God created Adam and Eve He created them with a Caucasian appearance. White people, despite being the keepers of the truth of the word of God more than any other race over the millennia, flatter ourselves if we think that God looks like us or created Adam and Eve that way. Most likely, Adam and Eve contained in their genes the traits of all races, and the branching off of their progeny into races occurred after Noah's flood (but that is getting ahead of the story!). So if a change in pigmentation wasn't the mark of Cain, what was? Keeping in mind that this is pure speculation, it seems reasonable to conclude that it was something akin to the "seal" with which God marks his followers in Revelation 7: 1-8. What that seal looked like or will look like is unknowable, but it could be something as simple as a tattoo on one's forehead that reads "Property of the Lord; Don't Touch!" Whatever it was, it must have been unmistakable, something that others would understand clearly, such that there would be no danger of anyone killing Cain out of ignorance. And whatever it was, we must assume that it was not passed down genetically from Cain to his descendents. But even if it had been, that genetic mutation was destroyed in Noah's flood.

Getting back to our earlier question now, "who were all these mysterious other people to whom both Cain and the Lord were referring? Cain's other brothers? His father? A whole other unnamed family tree of the human race? Or perhaps angels or fallen angels?" Since, as with so much of Genesis, there is no way to know, speculation and conjecture must suffice. First of all, in trying to answer these questions, we should remember that just because the Bible doesn't mention a thing doesn't mean that thing doesn't/didn't exist. It is completely possible that God created another race of beings apart from Adam and Eve. If so, they did not survive the flood, and it becomes somewhat a moot point what role they played in the drama of human history, although they may have left us some important archaeological material to brood over. Today, all human ancestry traces back to a single female—the one we call Eve. So, even if there was some group of proto-humans, semi-humans, or alien humanoids who roamed the earth millennia ago, none of us can be descended from them . . . at least not if we are to believe the story of Noah and the flood coming up in Genesis Chapters 6-10. (It is also possible that Cain was not referring to fellow humans at all but rather to carnivorous animals being the threat to his life. After all, he would now be alone in the wild. How would he be able to defend himself against a pride of lions or a pack of wolves? If so, the mark, whatever it was, would have kept those man-eaters away.)

Verse 16 takes us back to a topic covered in the previous chapter of our study—that of where the Lord put Adam and Eve when He drove them out of the garden. As mentioned then, we cannot know the location in relation to the garden; we can only know that the Lord had the eastern gate blocked off. That does not necessarily mean He put the couple on that side of the garden, but it seems to imply it. Regardless, we do know that when the Lord drove Cain out of his presence, He sent him to the land of Nod on the eastern side of Eden. Now, if we make the assumption that the "Garden" of Eden is a particular location within the larger region called "Eden," then it is possible that the Lord put Adam and Eve out on the eastern side of the garden, which was actually located on the western (or southern or northern) side of Eden. If so, there was obviously a lot of land east of the garden, and that may be where the Lord could typically be found entering and exiting the garden. If so, then that is where humans would have wanted to be if they were looking to be in his "presence." Most likely, the Lord resided in the garden, while Adam and Eve resided outside the garden but still right beside it—close enough to be frequently in the Lord's physical presence. Clearly, Adam and Eve, as well as any other children they may have had at the time, lived in the Lord's presence, wherever that was, as had Cain, prior to his banishment.

So the Lord drove Cain to the far edge of Eden, into a whole other region to the east. It was far enough from the rest of the family that Cain feared he would be cut off from any regular interaction with either the Lord or his parents and siblings. Yet it was close enough that he simultaneously feared that someone might eventually go there to hunt him down and kill him (that is, if we reject the wild animal hypothesis mentioned previously). Did Cain know and understand the implications of the distance between Nod and the garden? Did he know anything about the difficulty of travel between those places, not to mention the time such a migration eastward would take either for himself or for anyone who might come looking to kill him there? If so, how would he know? Had he been to Nod before? If so, how and why? Had the Lord taken him or any other human around, perhaps in an aircraft of some kind, just to show them what was out there? Had he farmed some of the land in that direction? Had he visited Abel who was herding sheep or cattle out there? If, by contrast, he didn't already know anything about the place to which he would be sent, why would he be so sure back in verse 14 that he would be hidden from the Lord? Did the Lord never visit Nod? How would Cain know that unless he was familiar with the area and/or the Lord's comings and goings?

These are unanswerable questions, of course, but they pale in comparison to the next question, as prompted by verse 17: who was Cain's wife? We must answer that question with a question—what are the possibilities? One, his sister; two, someone from a whole other race of beings that the Lord created on earth; and three, a mate that God created especially for him due to his extreme situation. Let's look at each briefly. The first is plausible and even probable. The immediate issue that arises in people's mind pertaining to it is that of incest. Those who would say it could not have been one of his sisters because the Lord surely would not allow an incestuous relationship like that would be wrong for several reasons. One reason is that the same dilemma would have faced Cain's other brother, Seth. Who was *his* wife? Rarely do we think to ask that question because we are fixated on Cain's story. Yet, every human, every descendent of Adam and Eve that has survived on this planet to the present, must be related. Obviously, it is only the degree of separation that prevents us from thinking of ourselves as engaging in incest. Therefore, unless God made a special partner for Seth, too, then he certainly married his sister. That is a good defense of the first possibility. If that is the case, though, it means that one of these sisters had the misfortune of being Cain's wife. ("Misfortune" because that whole branch of the family tree would eventually be obliterated in the flood.) Yet that would not make her any worse off than all of her other sisters and their descendents except one—the wife of Seth, for all of them, except one little subset of Seth's line, perished in the flood.

The second possibility is certainly plausible, and it could help account for some archaeological anomalies that appear to be dead-end branches in the evolution of homo sapiens. Neanderthals come immediately to mind. It does seem, of course, that the Bible would have mentioned if there were other humanoids on the planet along with Adam and Eve. After all, the Nephilim are mentioned, although they haven't entered the story as of Chapter 4, which proves that at some point there was at least one other group of humanoids on the planet besides pure homo sapiens. Could it be that Cain's wife was one of them? It would seem that even if she wasn't, her descendants were the ones who were the victims of the alien interbreeding that produced the Nephilim. That would stand to reason because Cain's family was cut off from the presence of the Lord, and we should assume, therefore, cut off from the Lord's protection as well, making them easy targets for more advanced beings to make into sex slaves. Of course, it is equally possible that these descendants of Cain and his wife were not victims at all but rather willing and even eager partners of those fallen angels. Before we jump to any conclusions about the Nephilim vis-à-vis Cain's line, we should realize that it is also possible that some of Seth's descendants could have been their progenitors. Or even one or more of Adam's other sons could have been the one(s) to propagate with the fallen angels, because, again, all humans and humanoids except 8 people descended from Seth (we believe from a literal reading of the Bible) were wiped out in the flood. Noah was certainly descended from Seth, and we assume his wife and son's wives were as well, but we don't know that for sure.

The third possibility seems the least likely. It does not stand to reason that the Lord would have gone out of his way to "create" another humanoid just for a man whom He had cursed. Some theologians would reject the notion of God creating anything after the first six days of Genesis Chapter 1, much less something as odd as a custom-made wife for Cain. If the Lord did, however, it could explain the same archaeological anomalies and theological mysteries mentioned above. There is direct evidence neither for it nor against it, but the seeming incongruence of such a special creation with the rest of the story makes this option the least appealing of the three.

Perhaps a more important question than who Cain's wife was is that of whom his sons married. As we are about to see, six generations of Cain's descendents are recorded, but nowhere does the narrator tell us where those wives came from. If incest was the means of propagating the species with Adam's children, would it not be likely that it occurred with Cain's children as well? After all, Cain and his wife (ostensibly his sister) were just one generation removed from Adam and Eve, so the compounding effects of incest that we see later in history or in current events would not have happened. The genetic degradation would have been kept to such a minimum, we must assume, as to be negligible. However, if the direct brother-sister incest continued through several generations, it could have resulted in severe degradation. In fact, that would potentially account for why the behavior of the species grew so wicked that God would decide to wipe out homo sapiens except for Noah's family.

Regardless of who Cain's wife was, the narrator tells us in verse 17 that she bore him a son named Enoch. This first son became the namesake for the first "city" mentioned in the Bible. Were archaeologists ever to find the remains of this first city, it would be perhaps the most important discovery of its kind in history. But how would they know if they did find such a city, since we have no information about it with which to work? If somehow it could be identified, it would, we must assume, be located in the land of Nod, which would help us come closer to pinpointing the location of Eden. How large of a city could it have been, considering the small population of the earth at the time? It probably would have been extremely small relative to anything we would call a "city" today. Perhaps all the narrator meant by calling the Enoch settlement a city was that it was a walled-in dwelling place that could accommodate several houses and a few public buildings. It was likely just an urban center of some kind, be it ever so humble, which served as a place of refuge and defense against wild beasts and Cain's brothers' children (should one of them come and try to kill him).

Starting with verse 17, we see that the list of names for Cain's descendents looks strikingly similar to the list of Seth's sons coming up in Chapter 5. Except for the inversion of certain names and the addition of "Seth" and "Enoch" on the second list, they could easily be considered identical. Consider the two lists side-by-side:

1. Cain, Enoch, Irad, Mehujael, Methushael, Lamech
2. Seth, Enosh, Cainan, Mahalelel, Jared, Enoch, Methuselah, Lamech

It is evident that the following pairings look similar enough as to be the same person: Cain and Cainan, Enoch and Enosh, Irad and Jared, Mehujael and Mahalelel, Methushael and Methusaleh, and, of course, Lamech and Lamech. All of the names except Lamech are spelled differently, but that should not be taken as proof that the two lists refer to different people. Arguably, of all the many little points we have covered so far in the first four chapters, this is the one that creates the most doubt about the authenticity of the story being told. It frankly seems that the narrator just mixed together the descendents of Seth and Cain. This is a difficult problem to explain away from an apologetics standpoint without simply being dismissive of it.

That problem becomes even more noticeable in light of what comes next in Chapter 4. Verses 20-22 tell us more about Cain's line. Lamech's sons Jabal and Jubal (twins, one must suppose) were the fathers of nomadic herding (and all it might entail as a way of life) and instrumental music, respectively, and his grandson Tubal-Cain was perhaps the first metallurgist. Yet we know, if we are to believe the flood story in Chapter 7, that all of Cain's line was ultimately destroyed such that there is no living descendent of Cain today. Why even mention Jabal, Jubal, and Tubal-Cain then, much less point out what they were/are known for historically? Moreover, the narrator does not refer to them in past tense as if they were destroyed in the flood. Instead, he refers to them as if they were the fathers of all those currently engaged in nomadic herding, musical instrumentation, and metallurgy. How confusing! What are we to make of all this?

To start trying to make sense of it all, let us assume a few things for argument's sake. Cain and his descendents lived in the land of Nod, but the land of Nod was not all that far from the Garden of Eden, nor from where his parents and siblings lived, which was on the outskirts of the garden. Let's say it was a day's journey by foot. Therefore, the different branches of the family tree would have been close enough to have fairly regular interaction with each other if and when they chose to. Cain and his line were not allowed to live with the rest of the family on the outskirts of the garden, but they were not entirely prohibited from interacting with the family. That interaction, however, would take place by other humans visiting Nod, not by Cain visiting them on their side of Eden. The interaction would have been scant at first—let's say for a few years—as no one knew for sure what type or how much of it the Lord would allow. Adam and Eve themselves, knowing full well the seriousness of disobeying the Lord, decided never to go to Nod. As much as it hurt them, they cautiously chose never to see their first-born son again. In time, however, their next son, Seth, was born. He grew up having never known his older brother but always hearing about him. Eventually, Seth wanted to meet him, so he decided to trek across Eden to Nod. By the time Seth got there, Cain, who was a generation older, had a son named Enoch who was his age. Meeting his nephew/peer Enoch, Seth was inspired to name his own son after him. Hence, he called him Enosh.

If this explanation is correct, it leads directly to another question: what accounts for the similarity of names thereafter? Did each generation of Seth's children think it good to simply copy the names of their cousins, uncles, or nephews? Or is it possible that, since all humans at that time spoke the same language, the names given to first-born sons followed a logical pattern? For a hypothetical example, let's use our imaginations. Pretend that "Enoch" and "Enosh" meant something along the lines of "Junior," "Irad" and "Jared" meant something like "My Son," "Methuselah" and "Methushael" meant "Looks like Dad" or something to that effect, etc. This possibility seems more plausible than the first. Bible scholars tell us, however, that these names actually have more interesting, and in some cases, cryptic meanings that seem to have nothing to do with one another. Irad, for instance, means something to the effect of "townsman," while Jared means "descending" or "going down." We could perhaps read into these names all kinds of speculative origins. Irad may have been born inside the walls of a city instead of out in a nomad's tent, while Jared may have been born when his mother went into labor walking down a hill to fetch a pail of water. It is impossible to know their precise meaning much less their origins, and again, we are forced to admit that these lists of names are troublesome from an apologetics standpoint.

That brings us back to the original issue of why the narrator brought up Jabal, Jubal, and Tubal-Cain whose descendents perished in the flood. Seth's descendents had interacted with them and become acquainted with the herding lifestyle, musical instrumentation, and metallurgy as a result. Seth's line had learned skills from them, in other words, that they would pass on to the rest of humanity through Noah and his sons. If so, this seems at first glance to be a peculiar assortment of skills. Why point out these particular skills then? Answer: reading beyond the text, what we can make of this list is that it covers broadly three main areas of civilized life on earth—farming, industry, and the arts. To be more descriptive, it covers animal husbandry, geological and chemical experimentation, and technological innovation, all of which raise the standard of living from the hunter-gatherer/dirt farmer/eek-out-a-living level to a more urban/civilized/leisure time level. Consider, for instance, that when we talk about animal husbandry as a skill that had to be developed or refined over generations, we are probably talking about humans learning to spin wool into yarn and weave yarn into fabric. With that skill, a plethora of modern conveniences emerged, such as comfortable and artistically

-fashioned garments, weather-proof shelters called tents, and the accompanying new businesses of manufacturing looms, clothes, and tents. Experimentation with making metal products likewise would have created new businesses, the most important of which no doubt were the manufacture of tools and weapons. It should go without saying that the development of art, whether in the form of music or something else, is a leading indicator that a society has progressed to a higher level of civilization, because its prerequisite is having a significant amount of leisure time.

Does this mean that none of the six generations of people in either Cain or Seth's line had made any advancements in these areas prior to Lamech's sons and grandsons coming along? No, probably not; that would be hard to imagine. It appears instead that humankind has been on a perpetual march of progress toward ever-greater degrees of civilization throughout history, notwithstanding that it has not happened uniformly but rather in phases of taking two steps forward and one step back. Keep in mind, however, that we are talking about prehistoric times here. As such, we must assume that life outside the garden was extremely difficult for a few generations. Consider that the Lord did not weave a nice set of wool underwear for Adam and Eve but rather presented them with, we must assume, a bloody animal hide to replace their hastily-made, itchy, non-durable fig leaf aprons! Later, as postulated previously herein, we think the Lord showed them how to make clothes out of animal skins. How many years did it take Adam or one of his descendents to figure out how to remove the hair from a sheep or goat in order to spin-and-weave it for clothing? From our modern vantage point, this seems like such common-sense technology as to be obvious. If we could go back to prehistoric times, however, when no tools yet existed for easy shearing, much less spinning and weaving, we would see how these early humans who lived in huts and hovels and who killed animals for food and clothing with sharp stones existed for several generations without what we consider obvious, common-sense technology. These early generations are likely responsible for giving us the caveman stereotype/myth and much of the archaeological material that is dated to the Paleolithic period.

THE FIRST POLYGAMIST

Moving on, verse 19 tells us that the last man in Cain's line, Lamech, took two wives. At first glance, this fact seems out of place; what does it have to do with anything? The answer: it probably indicates that, prior to the seventh generation, each man had only one wife. Therefore, Lamech must have been the first to break that tradition. We do not know whether the Lord had issued a command at some point before this saying each man must have only one wife. We do know, of course, that such was the Lord's will, because Christ says so (Mark 10: 6-9). Jesus corrected his questioners on this point, in fact, so straight-forwardly that it seems He was saying to them, "Duh! You know good and well that God intended every man to have only one wife!" It thus may have been understood from the beginning what the Lord's will was on the subject, but sin-prone humans began falling for the temptation to have more than one wife only in the seventh generation. Yet, after one man (Lamech) did it, many others assumed it was okay and began to emulate him. After all, the Lord didn't strike Lamech dead for doing it, and apparently didn't punish him otherwise. However, it is completely possible, even likely, that Lamech's action, like Adam and Eve's original sin (which also involved sex) carried natural consequences. Consider this: in verses 23-24 the narrator tells us that Lamech admitted to his wives that he had killed a man, but he doesn't say why. It is possible that this killing occurred because of a dispute with another man over the right to marry one of these women, either Adah or Zillah.

Using our imaginations again, let us picture that Lamech had already married Adah and had a conventional husband-wife relationship with her. Yet he was also attracted to Zillah, who was betrothed to another man, and his desire for her proved too great to resist. So he took her as his own and then had to fight the other man to keep her. This would be one explanation for why he confessed to his wives rather than to anyone else. The text doesn't say he kept it secret from others, but merely that he confessed it openly to his wives. Notice, too, that he immediately called to their attention the Lord's vow (made back in verse 15) to punish anyone who would harm Cain "sevenfold," and claimed for himself the right to be avenged ten times more should anyone seek to harm him. This indicates that Lamech was an unusually strong/self-willed individual, one who would go so far as to put words in the mouth of the Lord! The Lord had not made a promise to protect him (or at least if He did, the narrator did not record it), but Lamech claimed that protection anyway. Just as he had presumptuously taken a second wife without asking the Lord's permission, and just as he had killed the man who was to be her rightful husband (in the hypothetical scenario above), he presumes that the Lord is on his side and will not allow him to be harmed, and that he has a right to speak on the Lord's behalf. If this made-up scenario is even a half-way accurate reproduction of what took place in the real-life story of Lamech, is it any wonder that he and his sons represented the last two generations before the flood?

Assuming that something like this scenario took place, we must ask, did no one else try to stop Lamech from having this second wife? Where was his father Methushael? Did he not caution his son against it? If so, Lamech rejected his father's counsel. Did he have no brothers or cousins who tried to talk sense into him? If so, he didn't listen to them, either. Being the strong/self-willed man that he was, he was accustomed to getting his way, and so he did in the case of taking a second wife. Before we pile on Lamech too much, though, we should consider the fact that in Chapter 6, verses 1-2 tell us that over the course of time, the "sons of God . . . took wives for themselves of all whom they chose." This passage introduces the very real possibility that some non-human beings—presumably fallen angels—mated with human women in the era before the flood. There may be some other plausible interpretation of who the "sons of God" were, but in light of the overall explanation being presented herein, the presumption that they were fallen angels makes the most sense.

We will return to that topic of who the sons of God were later, but for now let's consider how the actions of the sons of God, regardless of who they were, affected Lamech and the men of his generation. It seems likely that they would have been jealous of these sons of God for picking the cream of the crop, so to speak, and leaving them with the less desirable females. They also likely would have been enraged if and when they saw one of their sisters taken by the sons of God. It is possible that this combination of anger and jealousy provoked Lamech to come to the conclusion that he had better claim and secure the females he wanted before one of the sons of God took them. Perhaps some of his fellows thought the same thing but were afraid to take action, fearing both the Lord and the sons of God, and likewise fearing disapproval from the family. Somebody had to be the pioneer. If so, Lamech, the most strong/self-willed man of his day, was that pioneer, the man who invented polygamy. The text says none of this, of course, and it is pure conjecture to propose it, but it does jibe well with the rest of *The Explanation*. We know that somebody somewhere at some point in history or prehistory invented polygamy, and this is as good an explanation as any other.

This raises another question that can be answered only with speculation—what did the Lord think about this change from his original plan of monogamous marriage for life? Using Christ's words again as our basis, we can assume that He didn't like it, but He also didn't judge it too harshly. On the one hand, we know that the flood came within one generation after polygamy was introduced, which indicates a possible correlation between those two events, but on the other hand, we know that even after the flood polygamy was reintroduced. This post-flood polygamy has, as of yet, not brought utter destruction like what happened in Noah's day. Therefore, we might do well to conclude that polygamy was not the main sin that provoked the wrath of God.

To move on toward the conclusion of Chapter 4 now, we see in verse 25 that Adam and Eve had a third son, whom they named Seth, only after Abel had been killed. There are a couple of points to be made here. First, the verse implies that a matter of years separated the birth of the first two sons, who seem to have been fairly close in age, and Seth, who was not born until his older brothers were either adults or at least teenagers and Adam was 130. If that is the case, does it mean that Adam and Eve had no sexual relations for all those years? No. It can't mean that, if we are to believe that Cain married his sister. That sister probably would have been at least a teenager, too, for it is hard to imagine the Lord allowing Cain to lead a prepubescent girl off into the wilderness with him. We cannot know the ages of these brothers and sisters, but we can speculate that perhaps Cain was, let's say, sixteen, and Abel was fifteen. The unnamed sister might have been fourteen. Of course, it is totally possible that Cain was more like sixty than sixteen, Abel was more like fifty than fifteen, and the unnamed sister was more like forty than fourteen. Either way, there could have been two, three, or more sisters born between the first girl and Seth.

The second point to be made about this verse is that the name "Seth" seems to mean something along the lines of "replacement." If so, the name is loaded with symbolism. We know that Seth's line survived the flood and therefore ended up replacing Cain's line as the purveyor of the species. We know, too, that Christ came through the line of Seth and that He took our place on the cross, replacing our sins with his righteousness. We also know that, considering how the first three humans to walk the face of the earth had all disobeyed God and proven themselves dishonest and untrustworthy, the Lord would have been perfectly justified to destroy the whole species right then and there rather than try to redeem it. This gives new and illuminated meaning to the opening phrase of John 3:16: "For God so loved the world ..."

THE FIRST PRAYERS

The very last verse of Chapter 4 is perhaps the most overlooked verse in the whole Bible, which is extremely unfortunate, because it is arguably the most important verse we have yet encountered in Genesis in terms of being applicable to our lives today. It says that once Seth's son Enosh was born, "then men began to call on the name of the Lord." What does this mean? Might it mean that prior to the generation of Enosh, people did not pray? If that is the case, why not? Did they not *know* to pray? The answer to this question is "No." Clearly, they knew to make sacrifices to the Lord, which takes more effort than praying. Why then would they not pray? Could it be because they did not *need* to pray? The answer is "That is correct." Why? Because during the first three human generations, the Lord was here on earth, or at least He visited the earth regularly in the flesh. We know this because, first, He was physically present in the garden when talking to Adam, Eve, and the serpent; and second, He was physically present outside the garden when talking to Cain and Abel. The text implies, therefore, that He was still physically present at least long enough to allow for the birth of Seth, probably long enough to see the birth of Enosh, but not long after—at least not regularly. If so, where did He go, how did He go, and why did He leave?

As far as where He went, we can easily visualize his going to "Heaven" (wherever that is) at this point in history. As far as how He went, it would be wise to remember that, while on earth, He was a physical being who had subjected himself to the limitations of the physical universe. Therefore, He probably would have left not by some miraculous disappearance but rather by a moving craft of some kind—whether moving through space, through time, or through dimensions, we cannot know. If either of the latter two options is how it happened, it may have appeared as though He vanished into thin air, as seems to have been the case with the risen Christ as recorded in John 20:19-29. As far as why He left, if we returned to the deistic ant farm theory here, we could speculate that He just wanted to see what we humans would do without his parental presence keeping order in the room. There is a better hypothesis, though. Having gotten his special creation established, surviving in the wild, and reproducing, He needed to leave them alone for three interconnecting reasons. One, the drama/plan of redemption had been drawn up before the foundation of the world was laid, and in order for it to begin playing out, He had to remove himself from the scene and only return to earth at certain specially appointed moments in history. Two, He wanted to give the humans maximum free will. To explain, as long as He was regularly visiting them in the flesh, they "knew" there was a God they had to answer to. Once the Lord stopped showing himself physically, however, the humans would have had to develop faith in his existence. Only by choosing to believe or disbelieve could they truly have free will in the sense that we understand it today. Three, the Lord knew it would take a special kind of dedication to continue believing in a deity that they could not see or hear physically. He wanted the fathers to teach their children to believe, but He knew that with each passing generation, it would become more difficult. Thus, over time, those who continued to believe would come to please him merely for having faith, and He would reward them accordingly at the end of the drama.

So what does it mean to "call on the name of the Lord?" Consider this: Adam and Eve apparently interacted with the Lord on earth on a daily basis, but Cain and Abel may have only gotten to interact with him sporadically, such as on occasions appointed for receiving sacrifices (whatever those occasions may have been). Now picture this: Cain moves to Nod, where he is out of fellowship with his parents and siblings and out of his proper fellowship with the Lord. Cain continues making sacrifices, and having learned his lesson, begins performing them correctly. Cain hopes to get back into favor with the Lord and consequently have his punishment rescinded so he can rejoin his family. The Lord does not reside in Nod but remains in the garden. He visits Nod periodically and receives Cain's properly-made sacrifices but does not rescind the punishment. Cain fears the Lord too much to stop making the sacrifices, yet he grows increasingly resentful over time, and his children grow up in that environment of sacrificial offerings made out of a resentful rather than a thankful heart. They inculcate that mentality and carry it over from generation to generation, and it consequently gets progressively worse.

Meanwhile, back on the other side of Eden on the outskirts of the garden, Seth is born, and he is many years younger than Cain. How many years younger, we cannot know. Adam has learned his lesson from all this, too, and he is careful to instruct Seth on how to make a proper sacrifice to the Lord. Seth, like Abel before him, does it correctly from the start. The Lord hangs around, visiting the outskirts year after year until He sees that Seth has had a son and has passed on the knowledge of proper sacrifices to the next generation. Satisfied that the pattern has been set, the Lord now departs, not to return until a specially appointed time in the future. Enosh, therefore, makes his sacrifices as his father taught him, yet the Lord does not show up to receive them! Nor does the Lord show up to receive Seth's, Adam's, Cain's, or Cain's children's anymore. The whole of the human race is aghast; what does it mean? "Has the Lord forsaken us?" they wonder. They begin to call out to the Lord. "Lord, where are you? What's going on? Did we do something to displease you? Lord, please talk to us! Please come back!" That's how men began calling on the name of the Lord, and it was the primitive beginning of a religious institution that we know today as prayer.

With that said, could it be that the Lord notified somebody—say Adam—before leaving, that He wouldn't be coming back anytime soon and telling him to man the fort, so to speak? Yes, probably so. He would have added, however, that He would be visiting from time-to-time, and that all humans should be vigilant about doing what He has instructed them to do and to avoid sinful things. In that sense, these first humans would be much like the church today. We are supposed to be looking for the Lord's coming at any moment, and we must be careful to be doing what He instructed us to do while avoiding sinful things. He has promised that He will come like a thief in the night, meaning suddenly and unexpectedly for sure, and some Christians believe secretly in the Rapture in the sky prior to his actual "second coming" to earth. He also said that just as it was in the days of Noah, so shall it be in that day. So what were the days of Noah like? That will be a main topic of the next chapter.

To conclude this chapter now, we should suppose that if humans began calling on the name of the Lord, it was because they honestly thought that by doing so they could persuade him to return to earth in the flesh and resume interacting with him like they formerly did. Since the Lord had always been in their midst prior to this, they had never considered that there would come a time when they would be left alone on the earth without his reassuring presence. One can easily picture Seth and Cain, if not Adam himself, feeling abandoned and lost, forlorn if not completely hopeless, scared and somewhat confused about this unexpected turn of events.

Moreover, we can visualize them standing out in the open field, with arms outstretched and with heads tilted upward, peering into the sky, wondering where exactly the Lord went and begging him to return. They may have seen him disappear into the clouds or some equivalent thing in the sky, and not knowing what lay beyond the blue sky and white clouds, thought He was "up there" just above them, perhaps looking down on them from close proximity. As they called out to him, therefore, they literally thought He was within earshot of their voices. By invoking him vocally, loudly, and in distress, to some extent they were foreshadowing Christ on the cross calling out, "God, why have you forsaken me?" As the years went by and the generations progressed, the children of Cain (and maybe Seth, too) probably began making their sacrificial altars bigger and bigger, rising higher and higher, hoping the increased size of the fire and smell of flesh cooking on an elevated platform would get his attention and bring him back. The problem with that line of thought and that course of action was twofold: one, the Lord had predetermined when He would come back, and no amount of conjuring on their part would bring him back early; and two, all that conjuring *did* catch the attention of the fallen angels, who were still very much present on earth even after the Lord left.

Those Satan-following angels, demonic in their hatred for God's special creation, had wanted to destroy the human race from the first day of Adam's life. They had been prohibited from doing so by the Lord thus far. Now that the Lord was physically absent, though, they knew their opportunity was here. They eagerly seized it. When men began calling out to the Lord, *they* began answering. A main reason the Lord had trained Adam and his children personally for all those many years is so they would recognize an imposter should one come into the picture posing as a "lord" or a "god." He, of course, knew in advance that was going to happen; He knew the humans as a whole would fall for the deception. Yet it was all part of the plan of redemption, and therefore it could have happened no other way. He also knew there would be a select few that would keep his ways even when everyone around them fell. And that will now take us into Genesis Chapters 5 through 7.

Genesis: The Explanation

What Really Happened
"In the Beginning"

5

THE EXPLANATION
GENESIS Chapters V - VII

AS IT WAS IN THE DAYS OF NOAH

Having gone slowly and methodically verse-by-verse through the first four chapters of Genesis, we will now shift into a higher gear and move rapidly but strategically through the next three chapters, following a topical approach rather than the earlier sequential approach. The first topic to discuss is genealogy. Biblical genealogies have engendered about as much fodder for skeptics and unbelievers as anything has, and Genesis Chapter 5 begins with one. Since we have already compared this genealogy of Seth's line with the list of Cain's descendents in Chapter 4, it would be redundant to say any more on that issue. The main points to be elucidated about the information in Chapter 5, therefore, pertains to the ages of these early humans, the mysterious "taking" of Enoch, and the reason for the Lord's choosing Noah and his sons as his special agents for preserving zoological life on earth.

> 1 This is the book of the genealogy of Adam. In the day that God created man, He made him in the likeness of God. 2 He created them male and female, and blessed them and called them Mankind in the day they were created. 3 And Adam lived one hundred and thirty years, and begot a son in his own likeness, after his image, and named him Seth. 4 After he begot Seth, the days of Adam were eight hundred years; and he had sons and daughters. 5 So all the days that Adam lived were nine hundred and thirty years; and he died. 6 Seth lived one hundred and five years, and begot Enosh. 7 After he begot Enosh, Seth lived eight hundred and seven years, and had sons and daughters. 8 So all the days of Seth were nine hundred and twelve years; and he died. 9 Enosh lived ninety years, and begot Cainan. 10 After he begot Cainan, Enosh lived eight hundred and fifteen years, and had sons and daughters. 11 So all the days of Enosh were nine hundred and five years; and he died. 12 Cainan lived seventy years, and begot Mahalalel. 13 After he begot Mahalalel, Cainan lived eight hundred and forty years, and had sons and daughters. 14 So all the days of Cainan were nine hundred and ten years; and he died. 15 Mahalalel lived sixty-five years, and begot Jared. 16 After he begot Jared, Mahalalel lived eight hundred and thirty years, and had sons and daughters. 17 So all the days of Mahalalel were eight hundred and ninety-five years; and he died. 18 Jared lived one hundred and sixty-two years, and begot Enoch. 19 After he begot Enoch, Jared lived eight hundred years, and had sons and daughters. 20 So all the days of Jared were nine hundred and sixty-two years; and he died. 21 Enoch lived sixty-five years, and begot Methuselah. 22 After he begot Methuselah,

> *Enoch walked with God three hundred years, and had sons and daughters. 23 So all the days of Enoch were three hundred and sixty-five years. 24 And Enoch walked with God; and he was not, for God took him. 25 Methuselah lived one hundred and eighty-seven years, and begot Lamech. 26 After he begot Lamech, Methuselah lived seven hundred and eighty-two years, and had sons and daughters. 27 So all the days of Methuselah were nine hundred and sixty-nine years; and he died. 28 Lamech lived one hundred and eighty-two years, and had a son. 29 And he called his name Noah, saying, "This one will comfort us concerning our work and the toil of our hands, because of the ground which the LORD has cursed." 30 After he begot Noah, Lamech lived five hundred and ninety-five years, and had sons and daughters. 31 So all the days of Lamech were seven hundred and seventy-seven years; and he died. 32 And Noah was five hundred years old, and Noah begot Shem, Ham, and Japheth. [NKJV]*

On the first point, it is obvious that the ages these men attained back then was vastly longer than what we enjoy today. Whether or not it was possible for people to live to be 900-plus years old back then is a subject of much controversy. Those who believe it was have the burden of trying to explain how. One theory is that, prior to the flood of Chapter 6, the composition of the earth's atmosphere was somehow different, thus allowing not only for humans to live longer and healthier, but also for some animals to grow larger; hence the size of some dinosaurs. Just as reptiles today continue growing right up to the day they die, so perhaps did dinosaurs in prehistoric times. One can easily imagine a typical lizard or a snake today growing to the size of a brontosaurus if it lived in the proper biosphere and had no natural predators. The same dynamic that produced the multi-centigenarian humans could have thus produced the extreme size of the

T-Rex, an animal which may have started out as nothing more than a flightless bird-type creature that we might envision as a meat-eating chicken. Given 900 or more years to continue growing, a small flightless lizard/bird without predators certainly could have bounded upward to 50 or more feet tall. Since no one can really say what different atmospheric composition would have produced that result, however, this theory is generally dismissed as being unlikely. (Later in this chapter, a theory will be proposed to show how it might not be all that far-fetched.)

Meanwhile, a different theory which has gained a lot of traction in recent years is that, prior to the flood, genetic degradation had not yet occurred to any great extent, so the bodies of humans (and perhaps all living things) were as healthy as they possibly could have been, with cells constantly being repaired or replaced to the highest degree possible. This would account for why there was no noticeable reduction in age during the generations from Adam to Noah. Jared (sixth generation), Methuselah (eighth generation), and Noah (tenth generation) all actually lived longer than Adam. To review the list of names and the ages attained:

- Adam = 930
- Seth = 912
- Enosh = 905
- Cainan = 910
- Mahalalel = 895
- Jared = 962
- Enoch = 365
- Methusaleh = 969
- Lamech = 777
- Noah = 950

Chapters 10 and 11 pick up the story of the descendants of Noah after the flood, and we see their ages begin to drop precipitously, just as the narrator tells us the Lord ordained when He decided to destroy zoological life on earth with a flood, capping the lifespan of humans at 120 years. We see that Shem was the last person to attain the age of 600. By the time we get to Abram a few generations later (at the end of Chapter 11), the normal age seems to have dropped to just over 200. Although this gradual reduction in age may comport with either of the above-mentioned theories, the latter is the one that seems to have the most direct bearing on the ability or lack thereof to attain an age measured in centuries rather than mere decades. Consider again that one reason the Lord never intended sexual coitus to be the means for procreating the human species is that He knew it would cause genetic degradation over time. Remember—he had intended Adam and Eve to live forever. If that is the case, would He have not also intended any/all other humans to live forever? The fact that, according to the New Testament, we will indeed all live forever in a new body in the future implies as much. But if so, why would there have been no noticeable reduction in the age of humans prior to the flood? Perhaps the aforementioned antediluvian atmospheric conditions had something to do with that. If that is the case, then what we see is that a combination of things caused the reduction in age after the flood. We will return to that topic shortly.

THE FIRST ASTRONAUT

Looking at the list of patriarchs cited above, Enoch is the wildcard on the list, living just over a third as long as some of the others—at least living on earth, that is, because we cannot know what became of him once God "took" him from the earth. According to verse 24, he did not die but was simply taken by God. Like so much else in Genesis, the narrator just throws this little episode out there in one verse with no explanation and leaves us guessing about the details. What does "and he was not, for God took him" mean? The conventional wisdom is that Enoch was transported off of the earth and taken to Heaven. As previously mentioned, where or what Heaven is cannot be determined. Some theologians have noted that this cannot be the correct interpretation of the verse, however, because it conflicts with John 3:13, which says, "No man has ascended to Heaven . . ." except Jesus. So he may have been taken off of this planet to some other indeterminate extraterrestrial location.

This episode, of course, adds credence to the ancient alien hypothesis, for someone or something not of this world apparently came down and picked up Enoch. There is a possibility that he was simply "teleported" (*Star Trek*-style) rather than "transported," or that he was taken through a time-travel process instead of a space-travel one. Regardless of the process employed, if the aforementioned parts of Genesis are to be believed, we should likewise believe that *something* unusual happened to Enoch. Whatever happened must have been something that others witnessed; else the narrator of Genesis would have reported that Enoch just disappeared rather than reporting that he was "taken." It is clear that he did not die, because the narrator tells us outright that all of the others on the list died. If Enoch had died, as some scholars suggest, why would the narrator have treated him differently? It is also important to note that he was not merely taken by some unspecified entity but specifically by God. As far as why God took him, the first part of verse 24 tells us it was because he "walked with God." From our modern Christian perspective, we understand that to mean he was an unusually holy, righteous man; in other words, he must have been morally superior and/or spiritually pure, steering clear of whatever common sins beset his contemporaries. We also tend to think he must have been, like David the psalmist would be famous for a couple of thousand years later, "a man after God's own heart" (I Samuel 13:14). Hebrews 11:5 says as much. There are other possibilities, however, for what caused Enoch to be singled out for the taking, and we should consider each one.

Secular ancient alien proponents, for instance, think of Enoch as a man of advanced intelligence, a scientist who somehow had figured out how to get the attention of the extraterrestrial "gods" who had "engineered" the human race. For whatever reason, these gods wanted Enoch to be with them (or perhaps to become one of them?). We cannot know whether this secular interpretation has any validity, but it would seem to comport somewhat with the story told in the apocryphal Book of Enoch. Although the Book of Enoch can be (and herein is) used as a supplementary source of information about many topics covered in the early chapters of Genesis, it is not considered scripture by most Christians and therefore should be understood more like Greek mythology than like the canonical Bible. That means it should be considered as a story based on a grain of truth which has been fictionally embellished. It is therefore of only marginal help in unraveling the mystery of what happened to Enoch.

Another possibility is one that weds elements of the Christian holiness theory and the secular scientist theories about Enoch: that he stood out among his fellows as the only one who had kept himself completely from being corrupted by the "bad" gods (i. e., the fallen angels/extraterrestrials).

This theory is related to the ancient Jewish idea of holiness, which can seem quite different from our modern Christian understanding of holiness: it is physical cleanliness, and even genetic purity, not spiritual (intangible/mystical) righteousness. Morality comes into play only insofar as it was considered "right" by God for homo sapiens to interbreed with homo sapiens and "wrong" for homo sapiens to interbreed with any other species, whether terrestrial or extraterrestrial, and Enoch was the first (perhaps only) one to understand that, or at least to grasp the implications of it fully.

To digress momentarily, one may ask, "What exactly was wrong with it?" The obvious answer is that the Lord (Yahweh) was the Elohim's chosen creator, and everything that He made was "good," and Satan and the fallen angels had created "bad" on the earth through genetic manipulation, which God never intended and would not long tolerate. A related but less obvious answer is that these modified species probably did not have "consciences" that allowed them to know right from wrong, and hence did not have "souls" that could be redeemed and made immortal. (We will return to this topic soon.) When God wiped them out in the flood, they were quite literally obliterated from existence, the same as we assume happens when any member of a lower-order animal species dies.

This raises an interesting side point. People generally do not think of ants, worms, snakes or other lower-order species as having souls or as otherwise being valuable enough to be resurrected from the dead, whereas we may think that dogs, cats, horses, or other higher-order animals are, especially if they are our own beloved pets. The question that arises from this point is whether our beloved pets go to "Heaven" when they die, and there is no way to know for sure. Testimonies given by many people who have had Near Death Experiences indicate that the place they went in their out-of-body experience was teeming with animals, at least the kind of animals we generally think of as the "good" ones. (No one comes back from Heaven complaining about the mosquitoes!) Beyond that, we can only speculate.

Getting back to the main point now, if it is true that Enoch was the one person on earth who really "got it" about not interbreeding with other species or with the supposed gods, why would God have removed him from earth? It seems that God should have wanted Enoch to teach his fellows this information—to evangelize, in effect. In keeping with the rest of *The Explanation*, however, we must assume that God did not want to short-circuit the plan of redemption by having a master teacher/evangelist purifying the human race with this information. Instead, He wanted to give humans maximum free will, and to do so, He had to leave the temptation (and ignorance) of people in place. If so, it is clear that Enoch had already passed this knowledge down to at least one of his sons, who likewise passed it down to at least one of his sons, and so on down to Noah and his three sons.

Now, with that said, it is important to note that just because one son managed to interbreed with a purely homo sapiens female for each generation from Enoch through Noah's sons does not mean that all the other sons of these patriarchs did. Nor does it mean that these patriarchs were exclusive in interbreeding with female homo sapiens, for it is totally possible that some of them were polygamists who had children with non-human or part-human females as well. All we know for sure is that at least one purely homo sapiens male-female pairing occurred in each generation from Enoch to Noah's sons, and that was enough to keep one genetic line pure through which humanity could be redeemed. We will have much more to say about all of this shortly.

THE FIRST SAVIOR

The third and final point to discuss from Chapter 5 pertains to the special calling the Lord placed on the tenth generation son of Adam called Noah. At some point after Noah's birth, his father Lamech pronounced that this son would become someone unique among all humans. How he knew that cannot be ascertained. In fact, whether he actually knew it or just made a statement of hope and optimism about Noah is a good question. In later chapters of Genesis, we see other patriarchs make similar pronouncements about their sons or grandsons, as if prophesying over them to forecast their futures. Noah himself did that for his sons and one grandson, for example, in Chapter 8, verses 24-27, and Jacob did similarly in Chapter 49. The nature of Lamech's pronouncement over Noah is interesting. It harkens back to the curse that the Lord had placed on the earth because of Adam's sin in Chapter 3. Lamech said, "This *one* will comfort us concerning our work and the toil of our hands, because of the ground which the Lord has cursed," but to paraphrase him, he basically said this guy is going to give us relief from our daily struggle for survival.

In the previous chapter of *The Explanation* in which this cursing of the ground because of Adam's sin was discussed, no mention was made of precisely how the Lord intended to curse the ground. Such a discussion is necessary now in order to show how Noah would bring relief from that curse. When the Lord said that the earth would produce "thorns and thistles," the implication was that those undesirable types of vegetation would grow in abundance naturally, but the desirable "herb of the field" would require intense cultivation. In other words, people would spend more time and effort weeding out the bad than planting and harvesting the good. Moreover, the soil itself might not provide the nutrients necessary for growing large, healthy edible vegetables and fruits. Perhaps it was too rocky, too sandy, too infertile to grow much of anything but weeds. Perhaps the earth even lacked a layer of topsoil at all. If so, how could that be? Keep in mind that, prior to the flood, the composition of the earth's atmosphere was probably different than it is today, and as we shall soon see, the composition of the earth's crust was also clearly different. If these components of the planet were different, why should it be hard to believe that the very surface of the earth itself had characteristics that we wouldn't recognize today as being normal as well?

It seems quite plausible that, before the great deluge, the surface of the earth resembled a moonscape, a mountaintop, or a desert plain, and that this surface sat on top of the earth like wrapping paper over a gift box. It did little more than hold back the "fountains of the great deep" (Genesis 7:11). If so, it most certainly was not spread uniformly over the whole earth, but rather had pockets of thick and thin layers. The thick layers were more fertile or at least more cultivable, but they were also scattered randomly, being few and far between. Meanwhile, the thin layers would function much the same as fault lines in earthquake-prone zones today. In other words, just like magma lies below the surface of a fault line and needs only for a tectonic plate to shift to send it upward through a crease (such as a volcano), so too did water percolate just below the surface of the pre-flood earth.

When people today think of the epic flood of Noah's time, they often think of it as having been caused primarily by rainfall. Genesis 7:11 does say that "the windows of heaven were opened," indicating that rainfall was definitely one contributing factor, but it also says that "the fountains of the great deep were broken up," indicating that subterranean water burst forth onto the surface. Of the two factors, it seems more likely that the bulk of the flood was caused by underground water springing up rather than by clouds in the sky raining down. If so, one point to ponder here is that of this subterranean water's temperature. People commonly overlook this issue when discussing the flood. But once considered, it is easy to see that, if the weight of the earth's crust bearing down on the layers beneath it is enough to generate heat that liquefies solid rock and metal into magma, all that weight probably would have had the same effect on the $H2O$ trapped below. To put it another way, it is possible that what sprang up from the earth when the "fountains of the great deep were broken up" was not water that was cool to the touch, but rather steam and boiling water. We should picture the flood, in fact, as a cataclysmic global steam bath rather than as an extended period of cool rainfall. Moreover, the steeping cauldron of steam that was the earth back then is probably what caused the break in the surface which released all that water in the first place. To paint another mental image, picture a pressure cooker boiling on the stove. As it starts to boil, steam begins to emerge from the safety valve on the cover. Having a safety valve allows the pressure to be maintained inside while letting enough steam escape to prevent an explosion. Now pretend the safety valve gets blocked off. In time, the pressure will blow the lid off; the cooker will explode. Likewise, lacking adequate ventilation, the earth's surface would have exploded.

Prior to this explosion, all that steam wafting into the atmosphere non-stop, day and night, year after year, would have made the earth like an ecological bio-dome. It would have spread the warmth rather evenly around the globe, keeping the poles from freezing over and preventing the equator from heating up to unlivable temperatures. Had the surface of the earth been uniformly composed of thick, rich topsoil, the planet would have been essentially perfect for nurturing life to the fullest extent possible in terms of size, health, and longevity. As it was, of course, the surface of the earth was unevenly divided between small pockets of rich topsoil and comparatively large areas of thin, infertile ground. The pockets of high quality land were undoubtedly coveted by humans, and perhaps by fallen angels, but certainly by a multitude of animal species as well. Before saying any more on this topic, we must pause and insert the text of Genesis 6 for reference.

> 1 Now it came to pass, when men began to multiply on the face of the earth, and daughters were born to them, 2 that the sons of God saw the daughters of men, that they were beautiful; and they took wives for themselves of all whom they chose. 3 And the LORD said, "My Spirit shall not strive with man forever, for he is indeed flesh; yet his days shall be one hundred and twenty years." 4 There were giants on the earth in those days, and also afterward, when the sons of God came in to the daughters of men and they bore children to them. Those were the mighty men who were of old, men of renown. 5 Then the LORD saw that the wickedness of man was great in the earth, and that every intent of the thoughts of his heart was only evil continually. 6 And the LORD was sorry that He had made man on the earth, and He was grieved in His heart. 7 So the LORD said, "I will destroy man whom I have created from the face of the earth, both man and beast, creeping

thing and birds of the air, for I am sorry that I have made them." 8 But Noah found grace in the eyes of the Lord. 9 This is the genealogy of Noah. Noah was a just man, perfect in his generations. Noah walked with God. 10 And Noah begot three sons: Shem, Ham, and Japheth. 11 The earth also was corrupt before God, and the earth was filled with violence. 12 So God looked upon the earth, and indeed it was corrupt; for all flesh had corrupted their way on the earth. 13 And God said to Noah, "The end of all flesh has come before Me, for the earth is filled with violence through them; and behold, I will destroy them with the earth. 14 Make yourself an ark of gopher wood; make rooms in the ark, and cover it inside and outside with pitch. 15 And this is how you shall make it: The length of the ark shall be three hundred cubits, its width fifty cubits, and its height thirty cubits. 16 You shall make a window for the ark, and you shall finish it to a cubit from above; and set the door of the ark in its side. You shall make it with lower, second, and third decks. 17 And behold, I Myself am bringing floodwaters on the earth, to destroy from under heaven all flesh in which is the breath of life; everything that is on the earth shall die. 18 But I will establish My covenant with you; and you shall go into the ark—you, your sons, your wife, and your sons' wives with you. 19 And of every living thing of all flesh you shall bring two of every sort into the ark, to keep them alive with you; they shall be male and female. 20 Of the birds after their kind, of animals after their kind, and of every creeping thing of the earth after its kind, two of every kind will come to you to keep them alive. 21 And you shall take for yourself of all food that is eaten, and you shall gather it to yourself; and it shall be food for you and for them." 22 Thus Noah did; according to all that God commanded him, so he did. [NKJV]

THE GREAT CORRUPTION

Over time, as the population of humans and animals alike increased, these highly fertile areas became the subject of violent territorial disputes. Starting with individual brothers, then familial clans, later tribes, and eventually kingdoms, humans fought and killed each other for the limited resources of the fertile lands. Meanwhile, the "gods" likewise split into warring factions and played one group of humans off against the next. Each group of these fallen angels sought to woo the strongest and smartest men to side with them and to woo the prettiest and otherwise most desirable women for recreation and procreation. The interbreeding created stronger bonds between each group of people and their respective "gods" during this prehistoric period just the same as happened all through recorded history when a conquering tribe or kingdom subdued a neighboring tribe or kingdom. The half-breed offspring became the intermediaries between the two groups, in this case the terrestrials and extraterrestrials. These children were the ones mentioned in Genesis 6:12 as being part of "all flesh" that had "corrupted its way upon the earth."

The fact that the narrator tells us that all *flesh*, not just all *human flesh*, was corrupt implies that the interbreeding agenda of the so-called gods was not limited to homo sapiens but included a variety of species, the offspring of which had not been intended by the Lord, and thus none of which could be allowed to exist for much longer. Therefore, when the Lord decided to destroy all but Noah's family in the flood, He also chose to save only those animal species He had created and intended for reproduction. If so, this raises a question: how is it that even after the flood, we see as part of recorded history examples of "gods" and hybrid beings featured? The most plausible answer is that, just because the Lord decided to destroy these unintended chimeras, griffins, and other Frankenstein-like creatures does not mean He prohibited their reintroduction. He merely promised after the flood never again to destroy the whole earth (meaning all the various animal life forms except a select few). Instead, from then on He chose to do it the other way around—to be selective in destroying these unwanted life forms rather than selective in saving the desired ones. How He did that is a mystery, but we can speculate about a couple of possibilities. First, it seems likely that the Lord introduced a genetic change into creation in which the hybrid animal species became incapable of sexual reproduction; that is, adult griffins could not make baby griffins, for example. Therefore, the fallen angels would have had to build them one at a time and repeat the process endlessly, which would have been too labor intensive and time consuming, hence too much trouble.

Second, it is possible that only after the flood did the Lord dispatch "good" angels to earth to fight against the fallen angels and keep them at bay. (We see no angels mentioned in Genesis before the flood, but we see them introduced into the story within a few generations thereafter.) Over time, the fallen angels were either vanquished or else forced to work so hard at trying to regain their stronghold over the earth as to make it not worth the effort. So they began to pull out, and they have been mostly relegated to the sky and space ever since (hence Satan gets labeled the "prince of the power of the air" in the New Testament) and to sparsely populated areas of the earth, including the bottom of the sea. Essentially, the Lord, and by extension, the good angels, have allowed them limited access to humanity and the rest of creation—not enough to bring about the need for God to destroy the whole kit-and-caboodle again, but enough to allow for free will and sin to continue, thus furthering the plan of redemption which He had set in motion before the first day of creation.

Moving on now, Noah is the pivotal figure in the whole Genesis narrative. He was the Lord's chosen one for preserving a pure homo sapiens species. He and his wife and sons and their spouses had descended from Enoch's line and had inherited the knowledge of genetic purity, which in turn kept their consciences intact, allowing them to know right from wrong and to feel shame and remorse for sin. By extension, they had souls that could be redeemed and thus made immortal. This is the grand explanation for why in earlier chapters of Genesis the narrator is specific in telling us that God created man in his "image" *and* his "likeness." The image refers to the genetic composition that produces a pure homo sapiens, and the likeness refers to humans having consciences and souls that can survive the death of the physical body. As far as whom his sons' wives were, we cannot know for certain from which parents or grandparents they descended, but we can assume that, as just mentioned, they descended through the line of Enoch. Now, if they were several cousins removed from Noah's sons, this relational distance would have mitigated any potential genetic defects from being introduced into the species which would cause the age, vitality, and intelligence of each generation on average to be less than the generation before it. Yet we know that the Lord decided to cap the age of humans after the flood at 120 years. But how? Perhaps He deliberately introduced a degrading genetic mutation by having Shem, Ham, and Japheth marry their sisters. To explore this possibility, consider that by Noah's generation, when the narrator tells us that all flesh had corrupted its way upon the earth, he meant literally that every human being outside of Noah's immediate family was infected with alien DNA. That would explain why the Lord chose to send the flood at that particular point in history as opposed to before that or after that—because if He didn't intervene immediately, there would be no more pure homo sapiens through which to salvage the species. Noah's sons would have been approaching the age of marriage, and having no pure-human females left to marry, the Lord must have instructed Noah to marry his sons to his daughters.

At first blush, this theory seems hard to believe, but upon further consideration, it becomes no more difficult to believe than anything else put forth so far in *The Explanation*. Consider this: even the most elementary of children's Sunday School lessons about Noah and the ark point out how Noah must have been the laughing stock of the antediluvian world. Year after year, and decade after decade, he continued building his giant floating hotel on dry land, while his neighbors lived in debauchery and poked fun at him. Where the children's version of this story may go wrong is in assuming that Noah was constantly evangelizing all the while, trying to convince his wicked neighbors to repent and join him on the boat. The Bible does not say that; it is likely just an unfounded projection of modern evangelical Christian values back onto an Old Testament patriarch. If the theory aforementioned about alien DNA corrupting the genetic makeup of zoological life on earth is correct, then the Lord would not have permitted those corrupted humanoids onto the ark, even if Noah had begged him to let them board. Moreover, if the main characteristic of "corrupted" flesh was its tendency toward violence as the narrator implies, then Noah and his sons may have spent just as much time defending themselves from the threat of physical assaults against themselves and their wives by their neighbors as they did in construction of the ark. If so, the very last thing that Noah would have wanted was those violent hybrids on the boat with himself and his family! Thus, if he was deathly afraid of them, he certainly did not want to save them from the flood just to give them opportunity thereafter to perpetuate the problem.

Before going any further, we must make a rather substantial digression to mull over a passage in the New Testament in which Jesus references Noah and the flood and consider its ramifications. In Matthew Chapter 24, none of the above is even implied, much less stated directly. To paraphrase it: "As it was in the days of Noah, so shall it be at the coming of the Son of Man. They were buying and selling, eating and drinking, and marrying and giving in marriage right up to the day when the flood came and swept them all away." The common way of reading these verses is to assume that these were just "normal" people going about their regular daily lives, and they were no different than any of us today. That may not be the point Jesus was making, or at least not the main point, however, because if the speculation above is correct, these so-called "people" were in no way "normal," meaning fully homo sapiens and thus having the "image" or likeness of God. If they were genetically corrupted, as already mentioned, they may have been soulless/conscienceless. This would have adversely affected their behavior to the extent of making them nothing more than sin machines— biological robots (picture them as something like zombies) that were incapable of anything other than following their animalistic appetites. They would have been incapable, for example, of empathy, altruism, loyalty, trustworthiness, or any such expressions of high moral character. More importantly, they would have been incapable of feeling shame, guilt, regret, or remorse when they sinned. In fact, they would have been adamantly opposed to any such qualities, because they could only relate to the works, character, and behavior of their "father" (Satan). Morality, therefore, would have been alien to them, and by living in rebellion to it, they would have felt instinctively "right," because they would be doing the will of their father. As such they would have been irredeemable.

Before dismissing this hypothesis as absurd, we should consider human interactions today in the non-Christian nations of the world. The values and sense of morality in these cultures are not the same as those of America or other nations with a Christian heritage. Although varying from country to country and tribe to tribe, every day billions of people buy and sell, marry, and go about daily life as high-functioning beings, existing within castes system that include slaves and untouchables, under oppressive forms of government such as communism and military dictatorships, in misogynistic cultures where the value of females and children is determined by the ruling men, and so on. Fundamentalist Christians and probably most other Evangelical Christians believe these same people who endure such hard lives here on earth will also die and spend eternity suffering unmercifully in hell if they don't hear and believe the Gospel. Now, if those who hold this theology are sure that God will sentence billions upon billions of people to such never-ending torment, why should it be hard to believe that He would sentence some unknowable number of hybrid humans who do not know him or his ways to die a one-time death in a flood? In other words, why do we think that He should have been willing to save these people in Noah's day?

Just because in the current church age He is not willing that any should perish but that all should come to know him? Well, we realize that thousands of them perish every day for lack of knowing him, and He just lets it happen. The point is, the Lord *wanted* those "people" in Noah's day to be killed off, not saved, for some reason, and no other reason seems to make sense other than that they were not fully human and thus were not made in his image and consequently were incapable of operating in his likeness.

Back to the topic of these doomed beings' animalistic behavior, the above hypothesis may give rise to an objection that some mammals, such as dogs and chimpanzees, among others, are capable of showing loyalty and sometimes of demonstrating other higher order behaviors. If so, they may actually exhibit more of the "likeness" of God than some humans today, and certainly more than the hybrid humanoids of Noah's day. This brings us back to the question we touched on earlier in this chapter of whether animals can be "saved" or redeemed for eternal life. As already mentioned, we cannot know for sure. The passage about wolves living at peace with lambs, and leopards laying down to sleep with goats, while cows, lions, and bears all eat straw side-by-side and toddlers play with cobras and adders (Isaiah 11:6-8) is one indication, however, that in the end God may redeem "all" of his creation, not just humans. Consider the wording of John 3:16: "For God so loved the *world* (not just the human species) . . ."

Revelation 22:15, meanwhile, tells us that dogs will be outside the gates of the New Jerusalem, alongside the humans (or humanoids?) who will have been condemned as liars, murderers, sorcerers, idolaters, etc. It seems doubtful that this usage of the term "dogs" is meant literally, however, because it makes no sense to single out just one animal species—and man's best friend at that!—for condemnation with wicked, unregenerate humans/humanoids. It seems more likely that this term should be interpreted the same way that Matthew 7:6 and 15:26 are commonly interpreted. Having already begun trudging through some deep theological weeds, now we are about to enter the Biblical jungle, so the reader is encouraged to follow the whole chain of thought that follows for the next few pages and wait to the end of it before rendering judgment about it. None of what follows is meant to be taken dogmatically; it is not asserted as fact. It is rather presented as *one* way to explain some otherwise unexplainable or difficult to explain theological issues. Here goes...

THE GREAT DIGRESSION

In Matthew 7:6, Jesus seems to be telling the people who will listen not to give what is holy to the "dogs," meaning ostensibly those who will *not* listen (believe and apply his teachings). The question is: how can those in the first group know who is in the second group until they first at least try to share Jesus' teachings with them? Perhaps the answer is found in Matthew 15:26 where Jesus seems to state matter-of-factly that the Jews are the "children" (of God) and the Canaanites (Gentiles) are "dogs." He lumps people together into two distinct groups, therefore, based on their physical lineage. What are we to make of this? According to the typical 21st Century American philosophy of political correctness, acceptance of all people regardless of their race, nationality, sexual orientation, etc., is "right" (morally superior), so Jesus' words in these verses seem discriminatory and thus unreasonable, if not completely outrageous and highly offensive. In fact, these words undoubtedly serve as a stumbling block and a rock of offense (Isaiah 8:14, Romans 9:33, and I Peter 2:8) to some people—individuals and whole ethnicities alike. This may account for why certain ethnic groups (mainly those of the Middle East who have chosen Islam as their religion) seem to have a predisposition for rejecting Christianity today. Their religious preference may be based on genetics, which is of course something they cannot control. The fact that the particular Canaanite woman in Matthew 15:26 received Christ's mercy based on her faith means that some "dogs" are redeemable, but it doesn't change the fact that, at least apparently, most are not.

Our common Christian conception of separating people into groups is, needless to say, based on people's actions and behavior rather than their bloodlines. To distinguish good *individuals* from bad *individuals* in terms of their actions and behavior is to separate the wheat from the tares (Matthew 13:24-30). This belief comports with what we see in the book of Acts, the Epistles of Paul, and basically the whole New Testament after the four Gospels. Orthodox Christian theology says, therefore, that Jesus was operating under and fulfilling the old covenant with the nation of Israel and the Jewish people in talking about Gentiles as dogs.

But if that is true, what are we to make of Matthew 25:31-46 in which Jesus discusses "sheep" *nations* and "goat" *nations* in the context of a future judgment? He seems to be referring literally to whole political entities called "nations" rather than individuals within those nations. He seems, therefore, to segregate nationalities into just two broad categories—sheep and goats—just like

He distinguished children (Jews) from dogs (Gentiles) earlier. If this hypothesis is correct, then Jesus was saying that some whole nations or people groups are "goats" or "dogs," and as such they are damned from birth (excepting a few individuals whose faith and/or compassionate deeds may lead them to Christ).

This is a hard pill to swallow for those of us who have been schooled in the common Christian theology that upholds John 3:16 as the paragon verse of our faith—God is not willing that ANY should perish (which we take to mean "go to Hell when they die") but that all should have eternal life (go to Heaven when they die). The main reason this notion of whole people groups being condemned is so difficult for us to accept is that we naturally assume that salvation and condemnation are and must be individualized. That is to say, on judgment day, each individual will stand before God and be judged one-by-one on his or her own merits. That may or may not be the case, however. God is not limited to handling things (situations or people) one at a time like we are. He is capable of lumping like-things together and dispatching them as a unit, and He may choose to operate that way on judgment day. Consider the wording of Revelation 21:24 which says in some translations "the nations of them which are saved. . .", which implies that there will be whole nations composed of them which are *not* saved.

Admittedly, this is a difficult concept to grasp, but largely it is because we are stuck on the one-sided theology of individuals being judged for either having the "likeness" (character) of God or not, and we fail to regard adequately the other side in which whole groups of people either possess the "image" (uncorrupted genetic composition) of God or not.

Consider this: just as Jesus seems to have lumped people groups together and stereotyped them, Paul later seems to do the same thing in reference to the Cretans in Titus 1:12. Even if, as some theologians believe, Paul was quoting someone else who said that and was urging Titus to rebuke him for saying it, this verse shows that it was prevalent in the early church that some professing Christians stereotyped others. We are taught nowadays from a very early age that stereotyping is wrong. Yet, all stereotypes are built upon same kernel of truth. Here the truth may well have been that the Canaanites, like all Gentiles in fact, were not fully human, having been corrupted genetically, and therefore did not possess the "image" of God. Only the Jews (not necessarily the Israelites of the defunct northern kingdom, either)—the only people group on earth who could trace their ancestry all the way back to Noah and even Adam—were fully human and thus had the image of God.

With that said, as already mentioned, Jesus was not locked in to condemning all non-Jews and consigning them to a spot in Hell, for once the Canaanite woman exhibited her extraordinary faith in his healing power, He gladly answered her prayer (granted her request for healing her daughter). By extension, we must assume that this type of faith would have made it possible for her, or any other non-Jewish "dog," to be redeemed for eternal life. Likewise, not all pure-bloodline Jews were "good" people, as evidenced by the Jews' rejection of Jesus as the Messiah and their rabid bloodlust to have him crucified (see Luke Chapter 23). They had the image of God, but not his likeness! The problem, however, may be that since the Jews who crucified Christ did not have the likeness of God, Christians today think that having his spiritual likeness is the only thing that matters, thus completely disregarding the importance of having the physical image of God. But consider that Paul takes great care to explain that God has not cast off his people (Romans 11:1), and it is clear that he is referring to the Jews (Romans 9:1-5). In fact, the main theme of the whole epistle to the Romans seems to be how Gentiles have been grafted in to the people of God. It logically follows that the Jews are therefore still the people of God. If so, they are his people in image only (excepting a few individuals), because the vast majority who are non-believers in Christ certainly do not possess his likeness.

This raises questions: if Jesus stereotyped people based on their bloodline, but differentiated between good people and bad people based on their faith despite their bloodline, what does that mean for us today? How are we to rightly divide the word of truth on this topic such that we have the proper application of it to our own real-life circumstances? First, if we admit that this concept of stereotyping whole groups of people is a (or perhaps *the*) stumbling block and rock of offense that the Bible speaks of, then by extension we should conclude that it is also the stone that Jesus spoke of that will either grind people to powder or break people who fall on it (Matthew 21:44). Undoubtedly, that means it will grind most of the "tare" individuals within these "goat" nations to powder, but the few who are willing to fall on it (to have their pride broken and accept Jesus' atoning sacrifice) will be spared. Second, we should realize that possessing the "image" of God (as they Jews did/do) will produce certain types of *physical* blessings/curses or salvations/condemnations in this present life but will not necessarily affect the Lord's final judgment of the person's eternal, spiritual resting place. (Consider Deuteronomy Chapter 28 as the scripture *par excellence* on this topic of blessings and curses coming upon people in the here and now as rewards and punishments). For example, God spared Noah and his whole family here in this physical world, but that did not automatically mean that all eight of them received spiritual salvation. (Consider 1 Corinthians Chapter 10 as evidence.)

As Christians, we assume that *physical* salvation of these Old Testament people means they were assured a spot in Heaven in the afterlife, but we can't actually know who will be saved and who will not on an individual basis like that. God is the judge, and only He knows whose names are written in the Lamb's Book of [Eternal] Life [in Heaven] (Revelation 21:27). We may look at certain of the Old Testament patriarchs—Noah, Abraham, and Moses, for example—and see no indication of any major wrong-doing on their part and predict we will see them in Heaven just as surely as we will see Peter and Paul from the New Testament. And we may look at others like Cain and Canaan and predict they will be in Hell along with Judas Iscariot because the only thing we know about them from scripture is the "bad" thing they did. But many other biblical characters are harder to predict. What about Lamech or King Saul from the Old Testament, or Nicodemus or Zacchaeus from the New? They have spotty records, or they fall into some kind of gray area. We are prone to think to ourselves without realizing it, "If I just had a little more information to work with, I could judge whether these guys will be saved or not!" Can we not see how foolish that is? We should no more think we can consign any of these biblical figures to a place in Heaven or Hell based on the scant information we have about them than we can judge our next door neighbor or the guy that runs the car wash down the street. Only God knows people's hearts. We should let him decide everyone's *spiritual* and eternal resting place. All we can do is look at the few actions or words of any of these folks which are recorded in the Bible and know what *physical* consequences followed them here on earth.

Moreover, Christians tend to read the story of Noah and the flood as saying that we had all better be "good" (exhibit the *likeness* of God in our character) today lest we, too, be swept away by the wrath of God like the humanoids of Noah's day. That is probably a correct interpretation, especially for those of us who are not Jews or full-blooded Jews, because most of us do not know who or what all is in our family ancestry, and we cannot claim to have the physical birthright that full-blooded Jews can. We can see that the physical birthright did nothing whatsoever to help gain favor with Jesus anyway. In fact, He specifically rebuked the unrighteous Pharisees for claiming Abraham as their father. He said the devil was actually their father. (Matthew 3:9 and John 8:38-40). Clearly, having the likeness of God easily trumps having the image of God in a head-to-head match, but would not having both be even better? Again, based on the central message of the book of Romans and many parts of the book of Revelation, we might do well to conclude yes. Bolstering this argument is the undeniable fact that almost all, if not all, the whole New Testament was authored by men who had both.

Noah is a prime example of a man who had both attributes—image and likeness—and God "chose" to save humanity through him. Noah was fully human, he had a soul/conscience, and God knew that he would faithfully follow his instructions. A person with a soul/conscience could do nothing else; he would either obey, or he would feel great shame and remorse for not obeying. It is not that a full human is always obedient; we have already seen how Adam, Eve, and Cain were all sinners. Yet, we also see that in the case of Adam and Eve, their sin was followed immediately by shame and regret. Cain's remorse is more difficult to see in the scripture, but we might suppose that he felt remorse only after being caught and hearing what his punishment would be. If so, it demonstrates a rather important point: even one generation removed from the divine creation of Adam, humans had already regressed into a more depraved state than what Adam and Eve's "original" sin had put humanity in.

Likewise, when Jesus "chose" his disciples, He chose those whom He could discern had souls/consciences, so He knew that, like Noah, they could be trusted to carry out his instructions. As He later put it, "My sheep know my voice, and another they will not follow." (Notice He didn't say "my *goats* know my voice!") How could He discern that, one may ask? Well, He was God in the flesh, after all. The God (Holy Spirit) in him had that power. It seems very much the same as "They Live!", a 1980's B-grade movie depicting a certain type of glasses that one could put on which allowed the wearer to see through the outer appearance to determine the difference between humans and alien look-alikes.

So to reiterate, clearly it is not enough to have the image of God in us; we must also display his likeness. While in Bible times in the Middle East it was certainly more likely that a person with God's image would also display his likeness, it was clearly *possible* that a "dog" or individual from a "goat" nation could display his likeness. When Jesus praised the Canaanite woman in Matthew 15:28 for her faith, He was saying in other words, "Woman, you have more of the likeness of my Father in you than most of the genetically-pure Israelites that my Father sent me here to rescue!" But she was the exception that proved the rule. A few other exceptions (a very small percentage of the whole population) have continued proving it ever since.

About having God's likeness, it is important to notice that Jesus never talked to any of his disciples one-on-one and tried to scare them into doing good or acting right. He chose them from among the crowd like sheep from among the herd while they were yet sinners (fishermen, tax collectors, etc.) and took them under his wing to teach them. He knew they would make mistakes, but He also knew He would not lose a single one of them to the wolf, except the one who was foreordained to betray him (John 6:39 and 18:9). The only people whom Jesus tried to scare were self-righteous religious folks, high-ranking government officials, and wealthy fat cats.

So, if Jesus never tried to scare his disciples, was He trying to scare us modern Christians with his comments about judgment in Noah's day in Matthew 24? If not, then what was the point? What was He getting at? The point may have been that, just as God had a predetermined, chosen way of saving homo sapiens in Noah's generation, so He would have a predetermined, chosen way to save the species again in this later generation. Like Noah, the followers of Christ have been chosen and predestined (elect according to the foreknowledge of God—I Peter 1:2); we have heard the warning, and we are to be prepared. Although we have a duty to try and save our fellows through missionary and evangelical activity (or so many Christians believe), we must not refuse to get on the proverbial boat ourselves while we are in the process of trying to convince some other person for whom we have great affection to get on the boat with us. Only the "chosen" or "elect" will be saved. Those who hear the message of the Gospel but refuse to comply with it do not have the "soul/conscience" necessary for salvation, which is, according to the hypothesis currently being described herein, actually a malfunction in their biological machinery caused by a genetic mutation.

If so, this raises the question, could God not rewire all people genetically in order to save them? Yes, of course He could, be as we have already seen throughout this study, if He did so, it would negate free will, and that is something He will *not* do. In effect, therefore, when someone chooses to accept Christ, repent, and try to live righteously, God rewires the individual at that time. Or, at least He begins the rewiring process, which is an ongoing work of the Holy Spirit inside the person. The fact is, God knows which people will allow themselves to be rewired and which ones will not, and that is how salvation is predestined for some and damnation predestined for others.

Like the subject of the Trinity, predestination is a point of theology which tends to tie even the greatest intellects in knots. It need not confuse us, however, if we begin by realizing that each individual's choices and behaviors are caused by his or her genetic composition. A person who is to too far genetically mutated from the original code God placed in Adam (or Noah?) *cannot* be redeemed only because that person *will not* accept redemption. That's how predestination works. If that is the case, one may ask, how do people get further and further removed genetically from the original code? The first answer, which seems obvious, is by generational degradation. Each generation will be a little worse off than the one before it, unless the process is arrested and reversed. This is why the Roman Empire was more brutal and oppressive than any of the earlier world empires, and why Jewish religious hypocrisy reached its zenith at the time that Jesus walked the earth. Both were cases of generational degradation, and it is why Jesus said that all the innocent blood shed from the time of Cain and Abel up to then would be charged to that generation (Luke 11:50).

This is also what is meant by a "generational curse," and it is why judgment against an individual can carry over to his children, grandchildren, and so on for several generations. It is not that the children and grandchildren have personally "done" anything to deserve judgment (yet), but that they inherited judgment (lack of protection from God against Satanic control or manipulation) because they inherited their father and grandfather's genetic degradation. Unless the process is arrested by divine intervention, there can be only one outcome for these people—eternal damnation.

In essence, the whole process began in the Garden of Eden, when Satan managed to insert one act of disobedience into the human condition. That one act quickly metastasized in the brains of Eve and Adam and caused other acts of disobedience to follow naturally. This corrupted gene pool was then spread into the first child born unto humans, and from there it simply grew worse with each generation in Cain's line. It was present in Seth's line, too, but not as far developed, owing to the hard lesson learned and the consequent willful rewiring of the brain (repentance) that had occurred in Adam and Eve over the time since Cain was born. The degradation of the human gene pool having already begun, a never-ending struggle took place in the minds of each person born in Seth's line, such that some kept their conscience/soul while others did not. Add to that the fact that the fallen angels continued to interact with these early humans and entice them to sexual perversion, and the eventual result was that only four human male-female combinations were left intact genetically by the time of Noah's generation. It did not matter, therefore, whether Noah *wanted* to see his fellow humanoids saved or not; they were predestined for destruction by God, who rejected them based on the number of mutations removed from his original model at which they existed. It is not that God was *willing* that they should perish, because that would clearly contradict scripture (John 3:16). But God knew (as only He could have known) that they *could* not be redeemed because they *would* not allow themselves to be redeemed. And that was because they were not fully human.

If so, this raises questions. Consider the hypothetical case of two siblings who share the same parents. Both are the same number of mutations removed from the original model (Adam). One turns out well. He is a hardworking, law-abiding citizen, a productive member of the community, and a church-going Christian. His brother is a drug-addict, deadbeat, and all-around ne'er-do-well. How can this be? The answer: the latter brother is not so far removed as to be irredeemable. He *can* be saved. Whether he *will* be is another matter, and only God knows. This seems to be the place that most of us are in when determining our spot in the spectrum of mutations. Most of us are neither absolutely pure nor absolutely corrupted but rather somewhere in between; hence most of us *can* be saved. What about a case in which two regular, normal parents raise a monster—a sadistic serial killer? How did we go from spiritually savable (like the parents) to unsalvageable (like the child) in one generation? The answer: we don't know if that's what actually happened. We just know there is a slippery slope, and not every person in any given generation slides down it at the same rate; perhaps some people hit bottom before others. Once on the bottom, though, they become unsalvageable. Is it possible then that, once on the bottom, two of these bottom-dwellers have a child that can be saved? Yes! With God all things are possible! Is it likely? No, unfortunately not. For every one that is saved (literally "rescued" or "plucked out"), there will be untold thousands or millions who won't be.

To a certain extent, we Christians are required to be callous and indifferent to the destruction of those who are too far removed from God's original design, as is implied in the parable of the wise and foolish virgins in Matthew Chapter 25. (This parable follows, not surprisingly, immediately after Jesus' discussion of "as it was in the days of Noah . . ." in Matthew Chapter 24.) In this parable, the Lord accepts five *wise* "virgins" into a special event and rejects five *foolish* "virgins." The fact that He calls *all* of them "virgins" is important, because it implies that they are genetically pure, or at least pure enough that they *could* be redeemed. So their behavior is chaste and just. None of them were totally depraved, living in abject debauchery, like the humanoids of Noah's generation. They were keeping themselves physically uncontaminated. However, the foolish ones did not *do* something they knew they were supposed to do, be doing, or already have done. They didn't put oil in their lamps in advance. What does this mean? It may very well mean that they were not actively engaged in building the proverbial "ark" that the Holy Spirit had instructed them to build. Consider this: it may be that every generation has its ark, and that every civilization has its ark. Select individuals (virgins, meaning those who are chosen by God) within each generation and civilization are instructed to help build that particular ark for that specific place and time. Some do it, and some do not. To take Jesus' example in this parable literally, half do and half do not.

Look, for an example, at the United States and Great Britain in the early 1800s. It seems clear in hindsight that the professing Christians within English-speaking civilization in that generation were called by the Lord and led of the Holy Spirit to save the world from the scourge of slavery. Some of them (abolitionists) obeyed and set out to do so, while others (indifferent or unconcerned individuals who were otherwise apparent Christians) resisted, saying in essence, "Slavery is no big deal; it's been around forever, so what's wrong with it?" This is tantamount to people in Noah's day saying, "What water? What rain? What flood? All I see is dry ground." These naysayers had the same opportunity as the doers of the Word—they possessed enough of the image and likeness of God that they could have repented and helped bring about the collective salvation of millions of people in the here-and-now, but they "slumbered and slept" (Matthew 25:5) through that brief window of opportunity. Meanwhile, the vehement opponents of abolition—the pro-slavery crowd—represented those "evil servants" mentioned in Matthew 24:48-49, who began to "smite" their "fellow servants" and "eat and drink with the drunken" (enjoy lives of pleasure at the expense of others). Those evil servants were caught by surprise and "cut . . . asunder" by the American Civil War of the 1860s. In other words, the Lord's time of "tarrying," waiting for them to repent, eventually came to a close just like the door on Noah's ark eventually had to close, and the judgment that was prepared for them was meted out. Likewise, at some point in the future, the invitation to the wedding feast of the Lamb will be closed and the door to the banquet hall will be shut, and this at the behest of the Lord himself.

The Lord whom we think of as infinitely merciful, who is not willing that any should perish, and who died to save as many as possible—that same Lord will order the invitation to the wedding feast closed and the door to the banquet hall shut. In that day, this same Lord whom we Christians know as our loving savior will be a ruthless judge and executioner of the unbelievers and the wicked. In that sense, "As it was in the days of Noah, so shall it be at the coming of the Son of Man."

THE GREAT PREPARATION

. . . Exiting the Biblical jungle now and getting back to Genesis, in Chapter 6 verses 17-22, we see God tell Noah that a flood of water is coming that will destroy all land animals, including birds. It is assumed that fish and sea creatures were not included in the plan of destruction. Trying to determine the fate of the amphibians, reptiles, insects, and other lower order animals in this plan is tempting, but it would be futile. It seems clear that the focus was on land mammals and birds, and the reason might be found upon acknowledging a few points: 1) they are more closely related to homo sapiens than are these other biological life forms, 2) they were the primary ones corrupted by genetic manipulation at the hands of the fallen angels, and 3) they could be not merely manipulated but actually inhabited by demonic spirits, as evidenced by the story of Jesus sending Legion into a herd of swine because they did not want to be sent into the "deep" (Luke 8:26-33) and by Jesus explaining that unclean spirits apparently inhabit "dry places" (Luke 11:24). The implication of this third point is that spirits can't live in any creature that doesn't have lungs and breathe air. If so, that still leaves open the possibility that they could potentially inhabit whales or dolphins or such. Legends of mermaids and mermen in pagan history indicate that fallen angels may have been engaged in genetic engineering of sea creatures, perhaps for the very purpose of making use of the vast deep. After all, the oceans do cover far more territory on earth than does the dry land, and it would be a shame to waste it if there were some way to make it useful.

So the Lord instructed Noah to build a giant floating zoo and home for himself and his family. He gave Noah specific instructions for how to do it: size, dimensions, and materials. Then He told Noah to bring the animals on board, to preserve them for reproductive purposes and for sacrifices. Skeptics of course have made about as much hay of this story as any in the whole Bible, saying it would be impossible to, first of all, *get* all the different species of birds and land animals of the whole earth to the ark, and second of all, *fit* all of them on the ark. Believers, needless to say, claim otherwise and have gone to great lengths to explain how both things could be possible. One common explanation for how Noah got all those different species to his location is that God brought them or led them there. This is not at all far-fetched if we consider that animals possess all kinds of instincts that humans don't have and cannot understand. How birds know to fly south for the winter, why some monarch butterflies fly from one (and only one) particular place in Mexico to one (and only one) particular place in Canada, and what causes salmon to swim upstream to spawn then die are just a few examples. It is not at all unreasonable to think that God can steer the brains of animals as if by remote control. Think about it: if humans living just a hundred years ago were to see a toy airplane being flown by remote control without seeing the electronic device that makes it go or the person manipulating the device, would they not think it magic? Would they not find it mysterious, shocking, and probably even terrifying at first? If so, why should the thought of an invisible God controlling the behavior of animals via an unseen force be any different?

Even if the animals migrated toward the ark out of instinct, the question is: how did those who inhabited the Americas and the isolated islands of the world cross the ocean to get to the Middle East? It seems that one of two things must have occurred: either God somehow had the animals transported from all the isolated places on earth where they originated (perhaps by sending angels to collect them), or the earth's land mass simply comprised a single super-continent at that time. Yet the former doesn't make sense, because it leads us to the untenable position that the angels therefore brought them across the ocean and deposited them on the shore of Africa, Europe, and/or Asia, whereupon instinct brought them the rest of the way. So the latter is the explanation that actually makes the most sense. The Americas and the islands were simply not yet broken off from the rest of the land mass. If that were the case, then all of the animals could have simply walked, flown, crept, or slithered toward the ark by instinct, without angelic transportation. We will return to this topic of the super-continent shortly.

A less-often-cited explanation for how Noah got all the animals to the ark, which seems only slightly more unbelievable, is that there weren't as many different "kinds" of animals back then as there are today. If we assume that the "kinds" theory proposed earlier in *The Explanation* is correct, then it is completely possible that only a few hundred species existed back then rather than the many thousands that are identifiable today. Secular scientists will, of course, laugh that notion to scorn because they are convinced that it took millions of years for all the different life forms to evolve. There is, however, just as much evidence against this belief as there is for it, and the secular experts simply don't accept that evidence because it doesn't bolster their case but instead provokes as-yet unanswerable questions. The "Cambrian Explosion" is, for example, thus far an unexplained phenomenon, which explodes (pardon the play on words) the notion that there should be multiple missing links between life forms found in lower strata and upper strata of the earth's surface. Even so, the fact that there are life forms at all in the geological record prior to the Cambrian Explosion seems to indicate that whatever happened to cause that sudden blossoming of so many different species was not "the" creation event mentioned in Genesis Chapters 1 and 2. We must remain open to the possibility that some outside force "punctuated" the earth to cause all that bio-diversity in the Cambrian period, whether a direct act of God or an indirect, natural act such as a major ecological change which sped the rate of evolution quickly and dramatically.

We will conclude this chapter of our study by inserting the text of Genesis Chapter 7. Having already covered the parts of it that are germane to this study, we will leave it to stand alone as a bridge to get us into a discussion of Genesis Chapters 8-11 coming up next.

> *1 Then the LORD said to Noah, "Come into the ark, you and all your household, because I have seen that you are righteous before Me in this generation. 2 You shall take with you seven each of every clean animal, a male and his female; two each of animals that are unclean, a male and his female; 3 also seven each of birds of the air, male and female, to keep the species alive on the face of all the earth. 4 For after seven more days I will cause it to rain on the earth forty days and forty nights, and I will destroy from the face of the earth all living things that I have made." 5 And Noah did according to all that the LORD commanded him. 6 Noah was six hundred years old when the floodwaters were on the earth. 7 So Noah, with his sons, his wife, and his sons' wives, went into the ark because of the waters of the flood. 8 Of clean animals, of animals that are unclean, of birds, and of everything that creeps on the earth, 9 two by two they went into the ark to Noah, male and female, as God had commanded Noah. 10 And it came to pass after seven days that the waters of the flood were on the earth. 11 In the six hundredth year of Noah's life, in the second month, the seventeenth day of the month, on that day all the fountains of the great deep were broken up, and the windows of heaven were opened. 12 And the rain was on the earth forty days and forty nights. 13 On the very same day Noah and Noah's sons, Shem, Ham, and Japheth, and Noah's wife and the three wives of his sons with them, entered the ark—14 they and every beast after its kind, all cattle after their kind, every creeping thing that creeps on the earth after its kind, and every bird after its kind, every bird of every sort. 15 And they went into the ark to Noah, two by*

two, of all flesh in which is the breath of life. 16 So those that entered, male and female of all flesh, went in as God had commanded him; and the LORD shut him in. 17 Now the flood was on the earth forty days. The waters increased and lifted up the ark, and it rose high above the earth. 18 The waters prevailed and greatly increased on the earth, and the ark moved about on the surface of the waters. 19 And the waters prevailed exceedingly on the earth, and all the high hills under the whole heaven were covered. 20 The waters prevailed fifteen cubits upward, and the mountains were covered. 21 And all flesh died that moved on the earth: birds and cattle and beasts and every creeping thing that creeps on the earth, and every man. 22 All in whose nostrils was the breath of the spirit of life, all that was on the dry land, died. 23 So He destroyed all living things which were on the face of the ground: both man and cattle, creeping thing and bird of the air. They were destroyed from the earth. Only Noah and those who were with him in the ark remained alive. 24 And the waters prevailed on the earth one hundred and fifty days. [NKJV]

Genesis: The Explanation

What Really Happened "In the Beginning"

6

THE EXPLANATION
GENESIS Chapters VIII - XI

OF RAINBOWS AND REGENERATION

Chapter 8 of Genesis marks a transition in the story. The waters subside, the ark comes to rest on a mountain, and all the people and animals eventually exit onto dry land and begin to repopulate the earth. Noah builds an altar and makes a sacrifice to the Lord, and the Lord makes a promise (to himself) never to destroy life on earth through a flood again. (Not until Chapter 9 verses 9-17 does the Lord inform Noah of this promise by explaining rainbows to him.)

> *1 God remembered Noah, all those alive, and all the animals with him in the ark. God sent a wind over the earth so that the waters receded. 2 The springs of the deep sea and the skies closed up. The skies held back the rain. 3 The waters receded gradually from the earth. After one hundred fifty days, the waters decreased; 4 and in the seventh month, on the seventeenth day, the ark came to rest on the Ararat*

mountains. 5 The waters decreased gradually until the tenth month, and on the first day of the tenth month the mountain peaks appeared. 6 After forty days, Noah opened the window of the ark that he had made. 7 He sent out a raven, and it flew back and forth until the waters over the entire earth had dried up. 8 Then he sent out a dove to see if the waters on all of the fertile land had subsided, 9 but the dove found no place to set its foot. It returned to him in the ark since waters still covered the entire earth. Noah stretched out his hand, took it, and brought it back into the ark. 10 He waited seven more days and sent the dove out from the ark again. 11 The dove came back to him in the evening, grasping a torn olive leaf in its beak. Then Noah knew that the waters were subsiding from the earth. 12 He waited seven more days and sent out the dove, but it didn't come back to him again. 13 In Noah's six hundred first year, on the first day of the first month, the waters dried up from the earth. Noah removed the ark's hatch and saw that the surface of the fertile land had dried up. 14 In the second month, on the seventeenth day, the earth was dry. 15 God spoke to Noah, 16 "Go out of the ark, you and your wife, your sons, and your sons' wives with you. 17 Bring out with you all the animals of every kind—birds, livestock, everything crawling on the ground—so that they may populate the earth, be fertile, and multiply on the earth." 18 So Noah went out of the ark with his sons, his wife, and his sons' wives. 19 All the animals, all the livestock, all the birds, and everything crawling on the ground, came out of the ark by their families. 20 Noah built an altar to the LORD. He took some of the clean large animals and some of the clean birds, and placed entirely burned offerings on the altar. 21 The LORD smelled the pleasing scent, and the LORD thought to himself, I will not curse the fertile land anymore because of human beings since the ideas of the human mind are evil

> *from their youth. I will never again destroy every living thing as I have done. 22 As long as the earth exists, seedtime and harvest, cold and hot, summer and autumn, day and night will not cease. [NKJV]*

Because the story of Noah's Ark is such a grand and famous episode of the Bible, it has been the subject of countless books, documentaries, movies, children's stories, and assorted other forms of entertainment and education. Consequently, it all too frequently draws attention away from other parts of Genesis, and the result is something akin to having a giant distraction suddenly appear in the midst of the larger story being told. While it is obviously an extremely important topic, for the purposes of *The Explanation* it is no more important than the topics before or after it. Rather than allow ourselves to be thrown off track by it, we should focus only on the part that is most germane to our study, because it will illuminate things to come in Chapter 9. Back in Genesis 7:7, we saw that Noah's three sons and their wives entered the ark, but none of the three as yet had any children. Then in Chapter 8 verse 18 we see the same three couples exit the Ark, still with no children. It seems curious that these sons, who were quite old even by the pre-flood era standard, had no children. The only suitable explanation for this anomaly is that either A) God had supernaturally closed the wombs of their wives in advance because He knew He was going to send the flood, or B) the Lord had warned the couples not to have children and thus to abstain from sexual intercourse because of the coming flood. Whichever is the case, He must have done so because either 1) He didn't want the women to have to go through pregnancy or weaning while on the boat, or 2) He didn't want them to have the burden of dealing with childrearing because of how time and energy consuming it would be during this few months when pure survival needed to be the only focus.

Now let us pick up the story with the restarting of the human race and civilization after the flood.

1 So God blessed Noah and his sons, and said to them: "Be fruitful and multiply, and fill the earth. 2 And the fear of you and the dread of you shall be on every beast of the earth, on every bird of the air, on all that move on the earth, and on all the fish of the sea. They are given into your hand. 3 Every moving thing that lives shall be food for you. I have given you all things, even as the green herbs. 4 But you shall not eat flesh with its life, that is, its blood. 5 Surely for your lifeblood I will demand a reckoning; from the hand of every beast I will require it, and from the hand of man. From the hand of every man's brother I will require the life of man. 6 "Whoever sheds man's blood, By man his blood shall be shed; For in the image of God He made man. 7 And as for you, be fruitful and multiply; Bring forth abundantly in the earth And multiply in it." 8 Then God spoke to Noah and to his sons with him, saying: 9 "And as for Me, behold, I establish My covenant with you and with your descendants after you, 10 and with every living creature that is with you: the birds, the cattle, and every beast of the earth with you, of all that go out of the ark, every beast of the earth. 11 Thus I establish My covenant with you: Never again shall all flesh be cut off by the waters of the flood; never again shall there be a flood to destroy the earth." 12 And God said: "This is the sign of the covenant which I make between Me and you, and every living creature that is with you, for perpetual generations: 13 I set My rainbow in the cloud, and it shall be for the sign of the covenant between Me and the earth. 14 It shall be, when I bring a cloud over the earth, that the rainbow shall be seen in the cloud; 15 and I will remember My

OF RAINBOWS AND REGENERATION

covenant which is between Me and you and every living creature of all flesh; the waters shall never again become a flood to destroy all flesh. 16 The rainbow shall be in the cloud, and I will look on it to remember the everlasting covenant between God and every living creature of all flesh that is on the earth." 17 And God said to Noah, "This is the sign of the covenant which I have established between Me and all flesh that is on the earth." 18 Now the sons of Noah who went out of the ark were Shem, Ham, and Japheth. And Ham was the father of Canaan. 19 These three were the sons of Noah, and from these the whole earth was populated. 20 And Noah began to be a farmer, and he planted a vineyard. 21 Then he drank of the wine and was drunk, and became uncovered in his tent. 22 And Ham, the father of Canaan, saw the nakedness of his father, and told his two brothers outside. 23 But Shem and Japheth took a garment, laid it on both their shoulders, and went backward and covered the nakedness of their father. Their faces were turned away, and they did not see their father's nakedness. 24 So Noah awoke from his wine, and knew what his younger son had done to him. 25 Then he said: "Cursed be Canaan; A servant of servants He shall be to his brethren." 26 And he said: "Blessed be the LORD, The God of Shem, And may Canaan be his servant. 27 May God enlarge Japheth, And may he dwell in the tents of Shem; And may Canaan be his servant." 28 And Noah lived after the flood three hundred and fifty years. 29 So all the days of Noah were nine hundred and fifty years; and he died. [NKJV]

Here a few comments will suffice. The Lord instructed the humans to do two things upon stepping onto dry land—one, eat animals, and two, engage in sexual reproduction. Both of these things had already happened prior to the flood, but neither had been explicitly required or encouraged by

God. Humans had merely taken it upon themselves to do both. Here the instructions are explicit, but they are still not very detailed. Therefore, we cannot know whether the Lord specifically told them to have sex *only* with fellow humans and to reject the sexual advances of any "gods" or other non-human creatures, but it seems obvious that He definitely intended that. We also cannot know whether verse 3 means literally that all non-human animal species *should* be used for food or merely *could* potentially be used for food. Genetically, these species would have been pure as opposed to corrupted, so we may well suppose that what the Lord was really telling them was, "If it tastes good to you, you can eat it. It won't hurt you." If so, the implication here is that eating genetically corrupted animals would have hurt them. Perhaps that had indeed been one of the contributing factors leading to the degradation of all those people who had perished in the flood—they ate genetically tainted meat and it caused birth defects or some such thing. This raises the question then, if all uncorrupted species were now fair game, why would the Lord turn around a few centuries later and prohibit the eating of certain animals in the Mosaic Law? To delve into that topic would be to digress unwisely from the topic at hand, so let's not do it.

EARTH DIVIDED, HUMANITY UNITED

Moving quickly toward the end of this study, we must trace the story of the repopulation of the earth after the flood, and that gets us back to the issue of the super-continent. Secular science tells us that all of the earth's land mass was indeed once connected, and that it was eventually split through the shifting of the tectonic plates in the crust. Biblically, there are two good possibilities for when and why it happened. First, as already mentioned, it seems plausible that the shifting of the plates is what caused Noah's flood. This would explain the meaning of the "fountains of the great deep" of Genesis 7:11 being broken up. Another explanation, however, appears in Genesis 10:25, which says that in the days of Eber, the earth was "divided." Eber, a descendent of Noah through Shem's line, is generally thought to be the one from whom the "Hebrew" name/identity is derived. The term "Hebrew," of course, is commonly used to describe God's chosen people prior to the time when Jacob, who was renamed Israel, entered into a special covenant with the Lord, which then cut other descendents of Eber (Hebrews) out of the story. The fact that the narrator is careful to tell us that Eber had two sons, Peleg and Joktan, and that Peleg was actually named for the dividing of the earth, indicates that this division was something of special importance. We must treat it accordingly and follow a lengthy train of thought to develop a theory for it that fits within the overarching explanation. Genesis Chapter 10 below will set up our discussion.

> *1 Now this is the genealogy of the sons of Noah: Shem, Ham, and Japheth. And sons were born to them after the flood. 2 The sons of Japheth were Gomer, Magog, Madai, Javan, Tubal, Meshech, and Tiras. 3 The sons of Gomer were Ashkenaz, Riphath, and Togarmah. 4 The sons of Javan were Elishah, Tarshish, Kittim, and Dodanim. 5 From these the coastland peoples of the Gentiles were separated into their lands, everyone according to his language, according to their families, into their nations. 6 The sons of Ham were Cush, Mizraim, Put, and Canaan. 7 The sons of Cush were Seba, Havilah, Sabtah, Raamah, and Sabtechah; and the sons of Raamah were Sheba and Dedan. 8 Cush begot Nimrod; he began to be a mighty one on the earth. 9 He was a mighty hunter before the LORD;*

therefore it is said, "Like Nimrod the mighty hunter before the LORD." 10 And the beginning of his kingdom was Babel, Erech, Accad, and Calneh, in the land of Shinar. 11 From that land he went to Assyria and built Nineveh, Rehoboth Ir, Calah, 12 and Resen between Nineveh and Calah (that is the principal city). 13 Mizraim begot Ludim, Anamim, Lehabim, Naphtuhim, 14 Pathrusim, and Casluhim (from whom came the Philistines and Caphtorim). 15 Canaan begot Sidon his firstborn, and Heth; 16 the Jebusite, the Amorite, and the Girgashite; 17 the Hivite, the Arkite, and the Sinite; 18 the Arvadite, the Zemarite, and the Hamathite. Afterward the families of the Canaanites were dispersed. 19 And the border of the Canaanites was from Sidon as you go toward Gerar, as far as Gaza; then as you go toward Sodom, Gomorrah, Admah, and Zeboiim, as far as Lasha. 20 These were the sons of Ham, according to their families, according to their languages, in their lands and in their nations. 21 And children were born also to Shem, the father of all the children of Eber, the brother of Japheth the elder. 22 The sons of Shem were Elam, Asshur, Arphaxad, Lud, and Aram. 23 The sons of Aram were Uz, Hul, Gether, and Mash. 24 Arphaxad begot Salah, and Salah begot Eber. 25 To Eber were born two sons: the name of one was Peleg, for in his days the earth was divided; and his brother's name was Joktan. 26 Joktan begot Almodad, Sheleph, Hazarmaveth, Jerah, 27 Hadoram, Uzal, Diklah, 28 Obal, Abimael, Sheba, 29 Ophir, Havilah, and Jobab. All these were the sons of Joktan. 30 And their dwelling place was from Mesha as you go toward Sephar, the mountain of the east. 31 These were the sons of Shem, according to their families, according to their languages, in their lands, according to their nations. 32 These were the families of the sons of Noah, according to their generations, in their nations;

and from these the nations were divided on the earth after the flood. [NKJV]

So what does the narrator mean when he says the earth was divided? One possibility is that he is referring to the scattering of the people after the Lord confounded their language at Babel in Genesis Chapter 11. Let us set the stage for exploring that possibility by inserting the text of that chapter here.

1 Now the whole earth had one language and one speech. 2 And it came to pass, as they journeyed from the east, that they found a plain in the land of Shinar, and they dwelt there. 3 Then they said to one another, "Come, let us make bricks and bake them thoroughly." They had brick for stone, and they had asphalt for mortar. 4 And they said, "Come, let us build ourselves a city, and a tower whose top is in the heavens; let us make a name for ourselves, lest we be scattered abroad over the face of the whole earth." 5 But the LORD came down to see the city and the tower which the sons of men had built. 6 And the LORD said, "Indeed the people are one and they all have one language, and this is what they begin to do; now nothing that they propose to do will be withheld from them. 7 Come, let Us go down and there confuse their language, that they may not understand one another's speech." 8 So the LORD scattered them abroad from there over the face of all the earth, and they ceased building the city. 9 Therefore its name is called Babel, because there the LORD confused the language of all the earth; and from there the LORD scattered them abroad over the face of all the earth. 10 This is the genealogy of Shem: Shem was one hundred years old, and begot Arphaxad two years after the flood. 11 After he begot Arphaxad, Shem lived five hundred years, and begot sons and daughters. 12 Arphaxad lived thirty-five years, and begot Salah. 13 After he begot

Salah, Arphaxad lived four hundred and three years, and begot sons and daughters. 14 Salah lived thirty years, and begot Eber. 15 After he begot Eber, Salah lived four hundred and three years, and begot sons and daughters. 16 Eber lived thirty-four years, and begot Peleg. 17 After he begot Peleg, Eber lived four hundred and thirty years, and begot sons and daughters. 18 Peleg lived thirty years, and begot Reu. 19 After he begot Reu, Peleg lived two hundred and nine years, and begot sons and daughters. 20 Reu lived thirty-two years, and begot Serug. 21 After he begot Serug, Reu lived two hundred and seven years, and begot sons and daughters. 22 Serug lived thirty years, and begot Nahor. 23 After he begot Nahor, Serug lived two hundred years, and begot sons and daughters. 24 Nahor lived twenty-nine years, and begot Terah. 25 After he begot Terah, Nahor lived one hundred and nineteen years, and begot sons and daughters. 26 Now Terah lived seventy years, and begot Abram, Nahor, and Haran. 27 This is the genealogy of Terah: Terah begot Abram, Nahor, and Haran. Haran begot Lot. 28 And Haran died before his father Terah in his native land, in Ur of the Chaldeans. 29 Then Abram and Nahor took wives: the name of Abram's wife was Sarai, and the name of Nahor's wife, Milcah, the daughter of Haran the father of Milcah and the father of Iscah. 30 But Sarai was barren; she had no child. 31 And Terah took his son Abram and his grandson Lot, the son of Haran, and his daughter-in-law Sarai, his son Abram's wife, and they went out with them from Ur of the Chaldeans to go to the land of Canaan; and they came to Haran and dwelt there. 32 So the days of Terah were two hundred and five years, and Terah died in Haran. [NKJV]

Babel was founded by Nimrod, the son of Cush and grandson of Ham. Nimrod was only the third generation from Noah. Eber was also the third generation from Noah, which makes Peleg the fourth. If we allow for a matter of years after the city is founded for the tower of Babel to be completed, it would make perfect timing for Peleg to be born right in the midst of the scattering that occurs in Chapter 11. Another theory is that the various family lines were at war with one another. That is certainly plausible, but it raises the question: what's so unusual about that as to lead a man to name his son for it? The answer, if the above theory is correct, is that it would have been the first time such a thing had happened since the flood. We know from Genesis 6:11 that prior to the flood, the earth was filled with violence. But there was no mention of violence after the flood leading up to the tower of Babel. Instead, we get the impression from Chapter 11 verses 1-4 that extreme unity characterized the family lines of Shem, Ham, and Japheth. They even reasoned among themselves that the purpose of building the tower was to reach up to heaven and make a name for themselves, "lest we be scattered abroad upon the face of the whole earth." They specifically determined *not* to be divided. Why? It seems that the answer is found in their method of avoiding division—to build a tower into the sky that would help them make a name for themselves. While scholars and laypeople alike have typically been more concerned with the reason for their building the tower, we are more concerned herein with why they would want or need to make a "name" for themselves.

We can dispense with the issue of why they thought they needed a tower summarily: because they feared another flood and because they wanted to get to the place in the heavens (sky) where they assumed the Lord and/or the other celestial beings lived, which they believed to be a place of safety. Let's look at these two items one at a time. First, why not just live in the mountains if they feared another flood? Probably because the fertile farm land was in the valley. Although some livestock can live in mountainous terrain, cattle and certain other livestock can't. If the cattle had to graze in the plain, and cattle were considered the best source of food, milk, and hides among all animals, where the best grazing land was probably trumped other considerations.

Second, if they were the only people on earth at the time, why would they need to make a name for themselves? Whom would they be hoping to impress or make notice them? The answer can only be the gods; that is, those who inhabited the heavens, watched over them, and sometimes visited them. The fact that the Lord did indeed come "down to see the city and the tower" in Genesis 11:5 indicates that the people had probably seen beings or spacecraft moving through and/or descending from the sky after the flood. Whether these beings or spacecraft were angelic or demonic, we cannot know. Most likely it was both. We know from Genesis 8:15 that "God spoke to Noah" while he was still on the ark and told him to evacuate it. Then in chapter 9 verse 1 we see God blessing Noah *and* his three sons, and in verse 8 we see God speaking to Noah *and* his three sons, which indicates that the Lord was physically present with them after the flood just as He had been with Adam, Eve, and Cain back in chapters 3 and 4. Whether the Lord routinely showed himself in the presence of men thereafter, we cannot know for sure, but it would be reasonable to assume that He did not. Just as He had given instructions to Adam that He expected him to teach his wife and children, so now did He give instructions to Noah, Shem, Ham, and Japheth, which He expected them to pass along to their children. Apparently, that first generation kept the rules and passed them along, as did the second generation. By the third generation, however, the amount of time that had elapsed since the last visitation of the Lord with humans perhaps created doubt about whether He was still up there somewhere. Perhaps they feared that his absence meant they had done something to displease him, and perhaps they had begun to question whether his promise not to destroy the earth again (Genesis 9:8-17) was still in effect.

The narrator does not tell us that the fallen angels were still visiting the earth and interacting with humans, but he doesn't tell us that they weren't, either. We might do well to assume they were. If Satan visited the garden and talked to Eve in Genesis Chapter 3, why should we think that he or his representative(s) did not visit with humans after the flood? If that were the case, then certainly these fallen angels would have been doing so to cause confusion and division and to persuade people to question God. If, therefore, the latest interaction between a celestial being and humans was instigated by Satan rather than the Lord, would not the people have questioned who was actually in charge up there in the heavens? Would they not think they should do something to get the "gods'" attention, to show them how strong and unified they were, lest these other-worldly beings try to harm them?

Keep in mind that Noah, who lived 350 years after the flood (Genesis 9:28), would have still been alive through the time of the building of the tower of Babel, the confounding of the languages, and the scattering of the families. He surely would have been a voice of reason, trying to tell his great grandchildren not to engage in such folly as building a tower. Shem, Ham, and Japheth probably would have echoed those sentiments. But the younger generation would have likely looked upon these old men as out of touch, and in Noah's case perhaps even as a senile alcoholic! Consider that the first misstep recorded after the flood is Noah's drunkenness, which led to his being naked and to Ham's sin of seeing it and telling his brothers about it. (Why, exactly, Noah and/or the narrator consider that such a wicked thing is of course fodder for wild speculation, but let us not concern ourselves with it here.) The point is that by the generation of Nimrod, humans had begun to think independently of their forefathers. They had not begun to think independently of one another, however, which seems odd, considering that there must have been a few hundred people on earth at the time, and in chapter 4 there were only four people on earth when the first murder occurred! What accounts for this anomaly? There is no way to know for sure, but it seems likely that they all made it the top priority to stay together as a "family." Perhaps they did so for security. They would have needed to protect themselves against the ravages of nature and the threat of attack by carnivorous animals. Besides, as long as Noah was still alive, they probably stayed together out of deference to him, despite seeing him as increasingly irrelevant to their other decisions.

What we know for sure, according to Genesis 11:2 is that they all settled together in the land of Shinar (which scholars tell us is ancient Mesopotamia) after having "journeyed from the east." Genesis 8:4 tells us that the ark came to rest "upon the mountains of Ararat." If that means Mount Ararat which is in Turkey today (and it is not certain that it does mean that), then Noah and his family probably departed the ark, headed in a westerly direction. That would have put them moving straight into the area of the headwaters of the Tigris River. Having crossed the river or having gone around the top of it, they must have then migrated south along it until reaching the valley between it and the Euphrates River. We also know from Genesis 10:10-11 that, in the same generation that Nimrod built Babel, he started the city of Accad. This aligns with what secular historians tell us about the first great kingdom in the Middle East, Akkad, which was located in the upper half of Mesopotamia. Likewise, during that same general time frame, give or take a few years or decades, the city of Nineveh was established, which was also located near the upper Tigris. So the site of Babel, which would have been the southernmost extent to which the family had traveled, was chosen, apparently by Nimrod, as the place to finally settle down.

Based on the genealogy of Genesis Chapter 11, all of this would have taken place within approximately 100 years after disembarking the ark. That wasn't enough time for the memory of the great flood not to be spoken of on an almost-daily basis, since Noah, Shem, Ham, and Japheth were still alive, but it was enough time for Nimrod's generation to begin thinking independently of those patriarchs. Another factor to consider in all this pertains to the animals. Since dangerous carnivores such as lions, tigers, and bears were repopulating the earth at this time, and since they reproduced at an exponentially faster rate than humans, staying close together would have given Noah's family their best chance to survive. This factor also accounts for why Nimrod was noteworthy among his brethren: he was the best hunter, implying not so much that he was the most skilled at killing wild game for food and pelts, but that he was the most skilled at knowing how to defend the family against attacks by carnivores. He may have been the one to reinvent better weapons than wooden spears, stone knives, and such, which was probably all these first post-flood humans had readily available prior to Nimrod's time. The knowledge of metallurgy, as we have already seen, was pioneered by a family line that died in the flood, but that does not mean Noah and his sons did not have access to those metal weapons and other objects on the boat. It does mean, however, that those things were in limited supply, and as the population increased, only a few people would have had access to them. Likewise, for the people to make more of them, access to the raw materials and opportunity to experiment with mining and smelting had to be rediscovered. Just because everyone would have agreed that more weapons were needed does not mean that everyone would have had the combination of curiosity, intelligence, and inclination to lead in that pursuit. Just as in the time before the flood, some people were naturally inclined toward farming, while others were inclined toward herding, hunting, fishing, weaving, construction, masonry, etc., so, too, after the flood people had their individually chosen occupations and preoccupations. Experimenting with metallurgy was undoubtedly a full-time preoccupation. Nimrod was probably the one who distinguished himself through some successful experiments.

Regardless, Nimrod's ability to defend the family would have given him power and influence over them. Besides, perhaps his own life-and-death struggles with wild beasts had led him to conclude that the family should not rely on God or "the gods" for protection. If the sky beings didn't protect them from animals, after all, why should humans think the gods would protect them from future floods? This must have been especially true in the generation since leaving the mountains and migrating into the Mesopotamian floodplain. Nimrod and family likely would have experienced an annual flood or two of the region already. Undoubtedly, they would have been spooked by it. With all this in mind, it is no wonder that they built walled cities and a tower. Whereas today we think of walled cities as protection from attacks by fellow humans, Noah's family saw them as protection from wild animals. The fact that they decided to build a tower in not just any city but the southernmost one in the floodplain should not be surprising.

HUMANITY DIVIDED, EARTH OVERSPREAD

Apparently, the Lord was not displeased with Noah's family for staying together and building cities for defense against the beasts of the field for a time, because He knew it was necessary to ensure the survival of the species He had made in his image. He did, however, expect the humans eventually to overspread the whole earth and subdue it, not just populate one little area of it, as evidenced by Genesis 9:1. Likewise He knew that eventually the carnivorous animals would have far more to fear from the humans than the humans had to fear from them, as evidenced by Genesis 9:2. By the time the family decided to build a tower, the Lord knew it was time to disperse the population. Besides, the building of the tower signaled a turning point in the humans' ability and/or willingness to walk with him by faith and not by sight. Why? Because the Lord had specifically promised not to destroy the earth with a flood again, and the tower represented the very embodiment of faithlessness and fearfulness. Moreover, the Lord surely knew his own plan was to discontinue interacting with humans in physical form going forward, and by extension, therefore, He knew the only sky beings that would respond to the prayers made from the top of the tower by showing up physically would be the fallen angels. The Lord was trying to steer the family away from seeking interaction with those devils, knowing they would be hard enough for humans to avoid already. The Lord, of course, would not interfere with their free will, though, so He used a minimalist approach to nudge people in a certain direction only ever-so slightly.

So the Lord confounded the language of humans. This should not be hard to believe for Christians, since we also believe that on the day of Pentecost, the original, miraculous speaking-in-tongues event occurred in Jerusalem (Acts Chapter 2). Precisely how or by what means the Lord did this in either case, we cannot know. Just as we don't know how many different languages were spoken in Jerusalem on the day of Pentecost, so we do not know how many different languages the Lord implanted into people at Babel. Speculation runs rampant on this point. Considering that there are more than 2,000 distinct languages spoken on the African continent alone today, it is common to assume that the Lord must have given every sub-unit within the larger family at Babel its own language. Our hypothesis herein, however, is that the Lord divided the people into just three languages—one for each of Noah's sons. This would have resulted in three basic groupings of people, two of whom would have begun migrating away from Mesopotamia immediately in opposite directions—one to the East, and one to the West. We will return to this migration momentarily.

The question is, what does any of this have to do with Peleg and the dividing of the earth? The answer is that Peleg was the cousin and contemporary of Nimrod. Whereas Nimrod was descended from Ham, Peleg was descended from Shem, and it would be through this family line, not Ham's, that God would bring Abraham, Isaac, Jacob, Judah, and Christ. Peleg was not the driving force in the story of the building of the tower, the confounding of the languages, and the dispersing of the people; Nimrod was, but the descendents of Peleg, not Nimrod, would be the subject of the entirety of the Bible story thereafter! Therefore, the narrator tells us that the earth was divided (meaning the whole population of the earth was dispersed) around the time Peleg was born. Curiously, after mentioning Peleg and his brother Joktan in Genesis 10:25, the narrator then proceeds to follow the genealogy of Joktan rather than Peleg for the remainder of that chapter, whereas he follows the genealogy of Peleg in Chapter 11 until arriving at the story of Abram. Why? The most likely reason to follow Joktan's line at all is because it represented the very last opportunity in the whole Bible to trace human history outside the genealogy of Christ. Consider: the human race branched into thousands of different tribes and nations, yet after Joktan, only one line can be traced all the way back to Noah and/or Adam, and it is Christ's, which runs through Peleg.

Another possible reason is that Joktan was the lesser of the two sons of Eber, and Eber is the supposed patriarch of the Hebrew people. The narrator introduces him before summarily dismissing him as being unimportant to the story, just as happens later in Genesis when the lesser of Abraham's sons (Ishmael) is discussed briefly before being dismissed from the story and when the lesser of Isaac's sons (Esau) is likewise discussed then dismissed. Obviously, three occurrences of this same type of thing in Genesis form a pattern. If so, might there be other instances in the Bible that continue the pattern? Yes, there are potentially several of them, depending on how we might want to categorize them. We could include, for instance, eleven out of the twelve sons of Israel in this category. Only the story of Joseph is told in any detail; the rest seem to be bit players. It is ironic that the story of Judah is not told in any detail, considering that it is his line that is followed up to the birth of Christ. Speaking of the birth of Christ, Jesus' earthly father Joseph is mentioned briefly in the Gospels before he, too, is summarily dismissed from the story. These are just a few examples. With such examples in mind, we would perhaps do well not to put too much emphasis on Joktan.

The narrator basically inverts the chronological order of Genesis Chapters 10 and 11 because, in 10:5 in the midst of the genealogy of Japheth, we see the "isles of the Gentiles divided in their lands; every one after his tongue, after their families, in their nations." It is interesting that he specifically uses the word "isles" as opposed to lands or regions or some such thing. Scholars typically think of this as referring mainly to the islands of the Mediterranean Sea between Turkey and Greece. Some scholars have even speculated that it refers to the British Isles. It seems more likely, however, that it refers to the largest area of islands on earth, which is in the Pacific Ocean. Indeed, moving due east from Mesopotamia, one eventually arrives on the coast of China. And directly out into the sea from China is Japan, which is one of the most notable sets of islands in the world. Our hypothesis herein, therefore, is that the descendents of Japheth moved east and northeast, fanning out as far north as Siberia and Mongolia, and as far south as Afghanistan, Nepal, and the northern region of India, but with one sub-unit holding the original family name of Japheth all the way to their final destination and namesake: Japan. The Japanese islands are, after all, the terminus of eastward migration through central Asia. Upon reaching Japan, there is nowhere left to go moving due east. Therefore, those who carried the Japheth family name deposited it there. It is not unlike how, later in secular history, Hungary is named for the wandering, conquering Huns, for instance, whereas France is named for the Germanic tribe called the Franks, and England is named for the Germanic tribe called the Angles.

One major objection to this theory is that etymological studies indicate that the Japanese people did not call their land "Japan" but rather "Nihon" or "Nippon." The term Japan apparently originated with the Italian explorer Marco Polo trying to translate what his Chinese hosts called the islands to the east, which was something along the lines of Cipangu or Giappan. Over time, this word became Japon in French in Japan in English. Considering the loose science (if it can be called a type of science at all) involved in the study of linguistics, we should not be overly concerned with such objections, because they amount to hair-splitting. Consider for one example just how inaccurate translations can be in terms of both pronunciation and spelling: one of the most famous, recognizable names in the English-speaking world is "Christopher Columbus." This legendary, historic figure never called himself that, because that is an English translation/spelling/pronunciation/corruption of his actual Italian name "Christoforo Colombo." Working for the government of Spain, his name as rendered by Spanish writers was "Cristobal Colon," which hardly resembles either his Italian name or its English derivation! But regardless of what He called himself at the time or what others have called him in their various languages, He is Christopher Columbus to us today who read, speak, and think in English. So, too, Japan is "Japan" because that's what we understand it to be in English, and Japheth is "Japtheth" only in English translations of the Bible! Were we to read the word as it was written in ancient Hebrew, it would read something like ׳ פֶת and would be pronounced something like "yefet." The point is, we could get mired in word studies like this to the extent that we are never able to make sense of what we're reading. One main objective of *The Explanation*, however, is to help us understand what we're reading in a way that has practical application. Thus the explanation of Japheth becoming Japan herein is meaningful and useful for our purpose.

Meanwhile, we know from secular history that at least some of the descendents of Ham moved westward and southward into the African continent, because one of Ham's sons, Cush, is recorded as having developed an empire in the region between Egypt and Ethiopia. Some scholars have suggested that the name which is translated into English as "Cush" should actually be "Kish," and they say the Kishites stayed in the Middle East. Regardless, we know that there was indeed a kingdom called Cush or Kush in Africa. We also know that some of Ham's descendents remained in the Middle East, as evidenced by the Amorites (meaning the "people of Ham") coming from his side of the family, and by the Canaanites settling along the eastern coast of the Mediterranean Sea (where most of the rest of the Bible story would play out). It seems likely that the earliest empire-builders all descended through the line of Ham, probably being responsible for Akkad in Mesopotamia and Egypt in Africa alike. It is possible that Ham's descendents actually fanned out in two opposite directions. While one group moved west and founded Egypt and the other kingdoms and tribes of Africa, another group moved east along the coast of the Indian Ocean, settling Pakistan, southern India, Malaysia and the rest of Southeast Asia, and even Australia. This would offer a plausible explanation for why Aboriginal Australians and certain other Southeast Asian peoples have many of the same physical traits as sub-Saharan Africans—dark skin, kinky hair, and wide noses—rather than the traits that are common among other Asians (who are descended from Japheth).

Whereas some scholars have speculated that the descendents of Shem were the ones who moved eastward toward China, our hypothesis herein is that Shem's line stayed partly in Mesopotamia and the surrounding region which we call the Middle East today while the rest moved in a northwesterly direction to overspread Europe. One reason to think this is because the Hebrews and later the Jews descended through the line of Shem, and we know they lived in the Middle East. Another reason is that some of these Hebrews and Jews were described as being "ruddy," meaning red-headed, light-skinned, and/or freckle-faced (King David), and/or "hairy" (Esau). The ruddiest and hairiest people on earth are those of European descent today. Africans, Asians, American Indians, and Pacific Islanders are not ruddy, and rarely are they as covered with body hair as white Europeans.

Although this line of reasoning may not be the most scientific way of arriving at a conclusion about the origin of different racial groups, it is arguably the most common-sense way. To follow a scientific approach would lead to looking at the human genome project and trying to trace DNA patterns. The problem with that is twofold: 1) the human genome project can trace only mitochondrial DNA, which means it follows female lines. Since all of the Bible except for the genealogy that leads to Mary, the mother of Jesus, follows male lines, it would produce a set of clues that would be contradictory of the Bible or at least quite confusing were we to trace the origin of racial groups by female DNA. 2) Perhaps more important is the fact that the human genome project traces all that DNA back to a single woman who lived in sub-Saharan Africa, and unless we want to revise our whole thesis about where the Garden of Eden was located, we cannot place Eve in sub-Saharan Africa. So again, this DNA evidence leads to a conclusion that either contradicts the Bible or at least fundamentally alters our understanding of these first few chapters of Genesis. And if altered too dramatically, not only does *The Explanation* fall apart, but unbelievers and skeptics get their unbelief or doubt reinforced. In time, some new technology might surface to bring DNA evidence into better alignment with Genesis. But for now, we must reject any conclusions resulting from the human genome project.

Genesis Chapter 11 concludes with a genealogy starting with Shem and ending with Abram. Chapter 12 begins with the story of Abram, and as such shifts the focus from the repopulation of the world after the flood to tracing a particular man's role in and contribution to history. Once the story of Abraham commences, the rest of the Bible concentrates on the little area that we call the Holy Land, Palestine, and/or Israel, with only occasional digressions to Mesopotamia, Egypt, Persia, Greece, Asia Minor, and Rome, and scarcely even a mention of any other areas. What had formerly been a global story, therefore, becomes a regional one. To non-believers, this regional focus generally comes across as exclusive, arrogant, and self-absorbed Jewish mythology. To believers, of course, it is the beginning of the second phase of God's preordained plan of redemption, which required an uncorrupted family line that could be traced all the way from Adam through Noah to Christ. Because the Bible takes such a radical turn at the introduction of Abram, this study will conclude here at the end of Chapter 11.

CONCLUSION

RECAPPING HOW IT ALL REALLY HAPPENED

Existing in a dimension apart from our own, outside of the observable physical universe, the Elohim is both an "It" and a "He" simultaneously. An ineffable being, the Elohim can best be understood as pure consciousness, awareness, energy, light, love, goodness, truth, and all-encompassing perfection. The Elohim is the "God" of Genesis Chapter 1, and "the Father" throughout the New Testament. It was He who spoke invisibly from above, saying, "This is my beloved son. . ." (Matthew 3:17 and 17:5). It was He to whom Jesus prayed, addressing him as "Father" (Matthew 6:9, 26:39-42), and it was He whom Jesus referenced on the cross as he quoted Psalm 22, saying, "My God, my God, why have you forsaken me?" The Elohim is personified and embodied in the form of Yahweh ("the Lord") starting in Genesis Chapter 2, as Jesus ("the Son") throughout most of the New Testament, and as the "Holy Spirit" now in the church age.

In our feeble human attempt to visualize that which is invisible and indescribable, the Elohim can be likened to a mother ship, which is both a "thing" (a living being) and a "place" (Heaven) at the same time. The Elohim cannot be contained *within* a place, because all proceeds from and must therefore be contained within Him/It. The Elohim radiates "glory" (Revelation 15:8, 21:11 and 23) which can best be understood as both putting out light and sending forth a feeling of pure goodness. It is this that John referred to when he said, "God is light, and in him is no darkness at all" (I John 1:5). When people who have Near Death Experiences describe going into the light, it is inside the Elohim that they enter (although this does not mean all people who die go to Heaven or that those who go there temporarily in NDE's are destined to spend eternity there; that is a whole other theological issue). To be "in" this light does not mean to be under the luminescence of a great light source, as would be the case with a person standing "in" the sunshine on earth, but it means instead to be completely drawn in and enveloped by pure light itself. It would be more accurate, therefore, to say that, when approaching the Elohim at the time of death, the light enters the soul rather than the soul enters the light.

From this ineffable entity that we visualize as a mother ship proceed all the detachable craft (heavenly beings), which are representatives of the Elohim that travel into space-time and appear to humans. These emissaries come and go as they receive instructions, are dispatched, and complete their assignments or missions. Operating according to a hierarchy, they range from Yahweh/Jesus at the top, to archangels, angels, cherubim, seraphim, and an indeterminate assortment of other creatures beneath them, each with his/its own unique assignment or mission. Yahweh's mission (so to speak) in the Old Testament was to walk on the earth, create man in his own image, supervise the affairs of this special creation in this unique place—including issuing instructions, prohibitions, and corrections to humans—and ultimately to choose a particular family line to carry uncorrupted DNA. Jesus' mission (again, so to speak) in the New Testament was to be conceived in and born of a human female, grow up and live in a human body, be tempted in all ways that humans are, never sin, allow himself to be put to death by sinners in order that He could offer himself as an atoning sacrifice on their behalf to the Elohim, and thus to redeem the special creation that He in the form of Yahweh made in his own image.

The reason this manifestation of the Elohim called Jesus had this "redemption mission" is because at some point in the indeterminate past after Genesis 1:1 but perhaps before the creation of Adam, one of the archangels (Lucifer) turned bad. Cast out of the mother ship (so to speak) and consigned to an existence within the confines of earth and its surrounding celestial territory, he, along with the innumerable angelic beings that followed his lead, has spent the rest of his time as "Satan," a being devoted to trying, with some success, to destroy the special creation by turning it bad like his angel-become-demon followers and himself. He could turn humans bad in two different ways—through physically corrupting the gene pool by interbreeding with them or through persuading them to disobey Yahweh. This is the story of the plan of redemption in a nutshell, and it is how the Bible, when read straight through from Genesis to Revelation, not only makes sense but in fact offers the best explanation for the mystery of existence.

Highlights of the story start with Satan or his representative the "serpent" talking to the first human female in the Bio-lab of Eden and bringing about the original sin—human sexual copulation—which produced degraded and ever-degradable DNA in the species. Banished from the protected environment of the Bio-lab and weakened by this degradation, each generation of humans descended further from the "likeness" of the Elohim; that is, further from goodness, love, and truth, which made them more susceptible to part two of Satan's agenda—interbreeding. Satan and his fallen angel/alien (non-human) minions had already spent untold thousands of years building technologically advanced civilizations all over the earth for their own benefit. The introduction of fallen humans into the wild after their banishment gave these alien beings playthings for their amusement. Presenting themselves as "gods" and showing themselves rightly to be bigger, stronger, and smarter than the humans, these fallen angels were easily able to convince most humans to do their bidding. Playing the role of the Annunaki, they enslaved humans, making beasts of burden of the men and concubines of the women. Some of their offspring became the Nephilim—giants with great physical strength and mental acumen who have been immortalized in pagan mythology as the lesser gods. These supposed gods, both the greater and lesser, were all selfish, spiteful, petty, and capricious, and they taught humans by their example to be like that, too. By the time of Noah's generation, all but one human family had unwittingly gone over to the dark side; all human DNA except for Noah's line had been corrupted through interbreeding and through the bad influence of the gods. This caused Yahweh to choose to destroy all these hybrid humanoids and in a sense to re-create man anew in his "image" and the Elohim's "likeness" through Noah and his sons.

All the hybrids were indeed drowned in the flood, but Satan and some of his demons were not. Some of them who had physical bodies perhaps lifted off the earth in spacecraft to await the subsiding of the water, whereupon they had to start over. Those who were disembodied spirits may have somehow hitched a ride as well. The fact that demons walk through dry places looking for a place to call home (Matthew 12:43 and Luke 11:24) and can apparently be drowned while inhabiting the bodies of animals (Luke 8:33) indicates the possibility that most of the demons must indeed have drowned, leaving a much smaller number of them to interact with humans thereafter. The remnants of their past civilizations were and are still evident as Stone Age megalithic sites and underwater ruins. But these satanic beings themselves could never again dominate humans through sheer numbers as they once could.

Meanwhile, if Yahweh could preserve the species intact physically, thus keeping his "image" alive, He could eventually restore in humans the "likeness" of the Elohim through the plan of redemption. To accomplish all this, He needed a single family line to remain uncorrupted for a couple thousand years until He could be dispatched personally into the gene pool in the form of Jesus. So He chose Abraham, Isaac, Jacob, Judah, and so on. He gave them increasingly more explicit instructions from one generation to the next, which were eventually codified in the Mosaic Law. All the taste not's, handle not's, and thou shalt not's were merely to keep order in the family, to make sure the enemy did not manage to woo them back into worshiping alien gods and interbreeding with demons. It worked, and in due time, Messiah was born through this family line. He accomplished his mission and made a way for all humankind to preserve the image and cultivate the likeness of God from then on. By accepting on faith his finished work on the cross and his resurrection, any human can now become what the Elohim intended before the beginning, and that of his or her own free will.

Genesis: The Explanation

What Really Happened
"In the Beginning"

BIBLIOGRAPHIC ESSAY

METHODS OF RESEARCH AND CITATION

The information in a book of this type must necessarily be drawn from a wide variety of sources taken from a smorgasbord of academic disciplines. It can only display much more width of coverage than depth. Those looking for an exhaustive list of books on the theology of Genesis, for example, will be disappointed, but so will those looking for an exhaustive list of any other genre of literature that comes into play in *The Explanation*, be it science, history, ancient alien theory, or otherwise. Instead, what readers will see at the end of this study in the "Bibliographical Essay on Select Sources Consulted" and the "Some Additional Sources Consulted" section is a smattering of this and that—hopefully enough to satisfy that I did my homework and did not just invent all of my opinions and speculations on this subject from nothing more than my imagination.

There are four things about my research methodology for *The Explanation* that are different from any other book

I've written and/or published. The first is that, after much deliberation, I decided not to use footnotes or endnotes in this one. Why? Frankly, it is because making individual citations noting where specific pieces of information in the text came from would have been so tedious as to take the joy out of this project, but also because they are unnecessary for lay readers who don't know or care about such things. Moreover, for this particular book, footnotes/endnotes would serve no great purpose beyond trying to impress fellow scholars and prove to them that I did serious research, and I have no interest in making such a pretentious display. I neither seek nor need the approval of academic specialists who know so much about so little as to cause them to exist within their own scholosphere where they speak their own language. Furthermore, such citations would be redundant of the information presented in the bibliographic essay, and would therefore be a superfluous and cumbersome addition (filler?) to an already-lengthy study. The same could be said about the second thing that is different about *The Explanation*: I chose to cite only the name of the publisher and the year of publication for each entry and to exclude the cities where the publishers are located. It seems unnecessary to tell readers where a publisher is located considering that such information can be found easily enough nowadays by an internet search, should anyone want or need to know it. After all, is the value of a book any greater because it was published in New York than in New Mexico? Is the value any less because it was published in Dayton than in London? The value lies in the information the book contains, not the location of its publication. The name of the publisher seems important, though, if only because publishers deserve some of either the credit or the blame for whatever books made their way into the libraries and bookstores where I found them. Dates of publication seem important, too, if only to show the reader how up-to-date or outdated any given source may be. A book that came out in 2008, for example, ought to be more relevant than one published in 1938, although there are exceptions.

The third thing that is different about *The Explanation* is that I began writing it first, then began scouting out sources for it later. I started writing this book, in other words, before I ever cracked open another book for the purpose of research. That was possible because I was using the vast storehouse of knowledge I had previously amassed over four decades of studying the Bible, as well as secular history, philosophy, science, and such. (One can pick up an amazing amount of what we might call "background information" on any subject if one sticks to it over a 40-year span of time!) I confess, too, that the History [Channel] television series *Ancient Aliens* played no small part in oiling the gears of my mind toward seeking to reconcile the competing claims of Judeo-Christianity, scientific naturalism, and alternative ontological theories. The accretion of information in my brain on the subject of ontology is one main reason I felt compelled to write *The Explanation*. Sometimes we all need to make a brain dump. The thinking goes like this: once I get all of these random thoughts organized and put into print, I can purge them from my mind to make room for a whole new load of information on some other subject.

Only after writing a complete rough draft of the first couple of chapters did I begin to consult library sources looking for specific information about various topics, sub-topics, and isolated points in order to bring my work into sharper focus. Having started that process, I began jotting down the books that follow, along with my thoughts on them. That leads to the fourth thing that is different about *The*

Explanation than any other book I've ever written: this one's list of sources is not alphabetical but rather topical. It is arranged in a way that seems most logical to me; it follows my stream of consciousness from topic to topic and from book to book within each topic. If the arrangement seems random, it is because I mostly read the books in random order and, therefore, acquired (what was for me) new information randomly. For that reason, whichever book is the "best" source of information on any given topic does not necessarily appear first in order here. Nor are the sources listed in chronological order of publication. In many cases, older books had already laid out a particular fact or argument, but a newer one fell into my hands first, and thus it became the one I seized upon.

Many books which I mention in the "Bibliographic Essay on Select Sources Consulted" as "see also" sources may well be better treatments of their respective topics, but they had the misfortune of being discovered after I had gleaned much the same information from a variety of other books. The "Some Additional Sources Consulted" section is reserved for books I read or skimmed that are fine sources of information, but which either fit in no particular topical category very well or which yielded no specific fact or insight that jumped off the page and grabbed my attention. Yet each added to my overall understanding of some sub-topic mentioned in *The Explanation*.

Readers may notice a dearth of articles, whether from scholarly journals, popular magazines, or news media in the bibliographic essay and additional sources list. It is a purposeful omission. I have an aversion to scholarly articles for two reasons: 1) they tend to be written in the jargon that only fellow scholars in the field can understand, and 2) they tend to be harder to gain access to than books and internet sources. That which is written in them, if important enough, will eventually make its way into a book anyway, so the only good time I can see to consult scholarly articles is when they are brand new and contain some kind of novel information or approach. About consulting popular magazines and news media: it seems to me that in this digital age, most of the same information can be found online, so there would be a lot of redundancy were I to cite those sources.

I frequently did quick internet searches looking for specific shards of information, such as Hebrew words, scientific terms, and cross-referenced Bible verses. I have chosen not to cite many online sources herein, however, for a few reasons that seem logical to me. One, the quality and reliability of internet sources varies greatly, and it is not my concern here to try to differentiate between all of them. A wealth of information can be found online pertaining to Genesis, and it ranges from the astute and profound to the utterly nonsensical. Scrutinizing online sources can be difficult (although certainly not impossible), and I figured my time would be better spent in other ways. Two, since they are readily available to anyone with an internet connection, and since they don't require any special ability to access (unlike old fashion library books), it seems superfluous to list them; readers can look up anything they want on the internet easily enough without being guided by me. Three, the internet contains so many sources, a large number of which I used, that to include all those www.whatevers would immediately double the length of this already bloated bibliography. Four, the genre of literature called Biblical Studies, especially the subfield that focuses on Genesis, is so incredibly vast that I can only provide a fraction of the printed books and articles here; it would somehow seem a grave injustice to scholarship to include online sources while omitting hundreds of worthy printed publications. And fifth, the nature of the internet is that some web pages are here today, gone tomorrow. This is, again, unlike old fashion hard copy, which will be stored away in a library somewhere until the crack of doom. So, honestly, the best way to handle online sources is to tell you up front that I looked at a bunch of them, I gleaned ideas from some of them, but I plagiarized none of them.

With that said, I must mention two particular online sources that were indispensable to my research: Biblehub.com and bestcommentaries.com/genesis. The former stands out as the best tool I have ever found for comparing and contrasting multiple translations of the Bible side-by-side. It allows for seeing the exact Hebrew or Greek terms used and their various meanings in English. It also has over twenty English translations, as well as several English commentaries to choose from. The latter lists over one hundred hard copy commentaries and ranks them according to user reviews. Since many commentaries say much the same thing, using the reviews of other researchers to differentiate between them was a time-saver.

Finally, it should be noted at the risk of redundancy that, having consulted literally thousands of sources over a 40-year time span, I find it frankly impossible to catalogue them all, mainly because it is impossible to remember them all! Since my interest in this subject began decades before my official "research" did, I honestly cannot say where I got some of this information from. The bibliography that appears at the end of this study, therefore, is merely a sample of the books that made enough of an impression on me as to be worthy of mention herein, and again, the ones I just happened to pick up first after beginning my "official" research on this project. The ones that taught me something new tend to be the ones that made the biggest impression.

I will conclude with two points: 1) a disclaimer offered by the authors of one of the science books I looked at and 2) a rationalization offered by the author of another. Carl D. Schlichting and Massimo Pigliucci, in *Phenotypic Evolution: A Reaction Norm Perspective* (Sinauer Associates, Inc., Publishers, 1998) made this epigraph-worthy quote, which I echo: "We started out by trying to read everything relevant, but the explosion of literature in many of the interconnected fields of inquiry covered by this book soon made it clear that such an ambition was naive" (xi).

Mary Jane West-Eberhard, meanwhile, in *Developmental Plasticity and Evolution* (Oxford University Press, 2003), explains that when she started her research in the 1980s, her field of study was relatively new. By the time she finished her book, however, the field had grown to the extent that it was now virtually impossible for one person to keep up with the literature, and if she had known that would be the case, she may never have started her research. Having started, she eventually finished by generalizing on many sub-topics because neither she nor any other individual scientist has the luxury of time to do primary research on every little thing that must be mentioned in a book of the magnitude of the one she was writing. She expressed fear that fellow scientists, who as a whole tend to be rather territorial with regard to their sub-specialty areas, would reject her work based on such generalizations. She defended her choice to trespass on their respective sacred territories by saying, "I can read." Having read what they, the experts, have written on those subjects, she felt justified in using it as evidence with which to make a point or draw conclusions. I wholeheartedly endorse that sentiment and ditto it in my own research.

SELECT SOURCES CONSULTED

GENESIS AS HISTORY VERSUS MYTH

In order to write a book like *The Explanation*, I had to confront a psychological hurdle the size of a mountain right from the start: whether to believe the Bible or not; more accurately, whether to take the creation story in Genesis seriously or not. When the shelves of college and university libraries are stocked with academic book after academic book that calls Genesis a myth, it gives one pause about treating it as sacrosanct. Each book I picked up that hammered home this theme that it is just a myth, no different than hundreds of other ancient Middle Eastern myths, tried my faith a little more. The first book I pulled from the shelf that fit this category was *Hebrew Myths: The Book of Genesis* (McGraw-Hill, 1964), by Robert Graves and Raphael Patai. It is a survey of ancient sources, some of which are fragments and others whole manuscripts, some of which are written in cuneiform and others in various conventional language formats, and some of which are well known and others obscure. It draws upon the Hebrew texts such as the *Talmud*, the *Midrash*, the *Apocrypha*, and the *Pseudopigrapha*, as well as Sumerian, Greek, Egyptian, and assorted other ancient gentile literature. It shows how certain Bible stories are much the same as what one would find in the writings of the gentile civilizations of the Middle East that predated the nation of Israel or were coterminous with it.

The most famous of these stories, of course, is the *Epic of Gilgamesh*, with its version of the great flood, the ark, and a Noah type character named Utnapishtim, of which I was well aware before reading this book. But there are many others that Graves and Patai brought to my attention as well. Examples include the creation of the universe and life on earth, the creation of Adam and Eve, the Garden of Eden, the role of the serpent, the origin of Satan, the Tree of Life, and the cursing of Cain and later of Ham, among many others. Each of these Genesis topics has a rough equivalent somewhere in ancient gentile literature, although the details are always different. The authors did not go beyond Genesis herein, so they had no need to mention the similarity between the Mosaic Law and Hammurabi's Code, but it would have bolstered their case to do so.

The point the authors make about these stories is that they did not originate with the Hebrews in the Torah. Most of the Old Testament, they say, was fairly well plagiarized from the neighbors and rivals of ancient Israel and was adapted to its political needs. Even the Hebrew name for God, Elohim, did not originate with the Hebrews, say Graves and Patai. It looks and sounds too much like the name of the gods of the Assyrians, Babylonians, Hittites, and Phoenicians, all of whom borrowed heavily from one another's culture. Not surprisingly, the authors do not take Genesis seriously, much less literally as actual history, and their dismissal of it as holy scripture is evident throughout. In the same vein is David Maclagan's, *Creation Myths: Man's Introduction to the World* (Thames and Hudson, 1977), which equates the words "In the Beginning" with "Once upon a Time." For me, as mentioned, this type of writing posed a psychological problem. Getting around this mental obstacle was difficult but not impossible. Had it proven impossible, of course, then you would not be reading *The Explanation* right now. I would have killed the project and moved on to some other topic. Instead, I faced the problem head on, meeting the intellectual challenge and incorporating arguments and/or apologies into *The Explanation* to answer it.

Another book that confronts the same issues but does so from a Christian ontological perspective is *Genesis Regained* (Sheed and Ward, 1969), by F. J. Sheed. Its theme is that, while similarities abound between Genesis and ancient pagan myths, the differences between them are even more numerous and notable. To quote Sheed, "In the study of Comparative Religion men can deceive themselves by listing elements of similarity without sufficient reference to the totality in which these elements have their existence, from which they draw their meaning" (37); and "In particular it is a temptation for scholars to pounce upon verbal resemblances and build towering constructions upon them" (55). He explains that the Genesis story is similar to pagan mythology in the same way that humans and other primates are similar. Elements of each can be found in the other, but that by no means makes them interchangeable, as if the value of one is no greater than the value of the other.

Example: in the Sumerian flood story, there were 9 "kings" prior to the deluge, whereas in Genesis there were 10 "patriarchs" prior to it. Another example: whereas the ages these patriarchs attained in Genesis stretches the imagination--969 years for Methuselah--they are a drop in the bucket compared to the oldest Sumerian; he lived a whopping 65,000 years! A third example: even though the Babylonian god Marduk had many of the same descriptors as Israel's Yahweh, he was not all-powerful; he shared his pantheon with at least 3,000 other Babylonian gods, and they are the ones who *granted* him control over the universe! A fourth example: although the Feast of Tabernacles was held on the same day as the Canaanites' main religious festival, the former, unlike the latter, did not include ritual sodomy, bestiality, and human sacrifice! (And speaking of sex, the emphasis that all pagan religions placed upon it in contrast to the downplaying of it in Genesis may be the most important example of just how different the Judeo-Christian religion is than them.)

Sheed makes clear that he believes Genesis to be a man-made document, not something written by the finger of God. He does not see that as contradicting the orthodox Christian belief that it, like the rest of the Bible, is the "inspired" word of God. He questions what constitutes "inspiration," not seeing it necessarily as some direct, divine revelation of information, but wondering if it is rather like having a hunch or following intuition. If that is the case, and the writers of the Bible were acting on nothing more than what any of us act on in daily life when deciding the right course to take, only time will tell whether or not the hunch/intuition was inspired by the Holy Spirit. If it passes the test of time, as the book of Genesis has, such that millions of modern people can still have confidence that its basic message is from God (even if many others don't believe it), then it is inspired. By this logic, ancient pagan mythology by contrast is exposed as just that—mythology—because no one believes a word of it today; no one, for instance, thinks that Zeus is the guy who's really in charge up there! (Do they??)

The illuminating points that Sheed makes supporting the belief that Genesis is inspired and that we should therefore take it seriously (not literally, but seriously) are too numerous to list here. A couple of examples will suffice. One is that Christ himself seems to have taken the book of Genesis seriously, and in some places even literally, such as when he discussed God's original intent about marriage in Matthew chapter 19, quoting Genesis 2:24 and expounding upon it as a fundamentalist Christian might be wont to do today. Another is an answer to the charge that Genesis chapters 1 and 2 contradict each other; no, says Sheed, they actually complement each other because chapter 2 tells what humans are made *of* (dust), while chapter 1 tells what we made *like* (God).

SECULAR SCIENCE VERSUS RELIGION

The next hurdle I confronted was that of science; specifically, the opposing points of view of modern secular scientists about the origins of the universe and life, the age of rocks, etc. I was prepared for certain now-standard arguments for why Genesis cannot be taken literally and thus should not be taken seriously, and I will get to them shortly. To start, however, Mano Singham's *Quest for Truth: Scientific Progress and Religious Beliefs* (Phi Delta Kappa Educational Foundation, 2000) proved most helpful in laying the mental groundwork for me. A foreign-born physicist who never identifies his own religious beliefs, Singham offers the most balanced account of the creation v. evolution debate, which he says should really be called the "creation v. naturalism" debate, than I've ever seen or likely ever will see. A litmus test, in my mind, for how fair a writer is to both sides is whether he leaves me wondering which side he agrees with, and that is the case here. On one page Singham seems sympathetic to the Genesis believers; on the next he seems squarely in the secular scientist camp. He approaches the subject like a debate coach—emotionally detached from either side's arguments. He summarizes the history of the legal cases involving the teaching of creation v. evolution in American public schools, and shows how the verdict in the court of public opinion is and always has been more important ultimately than any judicial verdict.

Singham explains that the root of the problem which causes so much confusion and thus keeps stirring up the issue from generation to generation is the lack of clear definitions of what "science" and "religion" actually are as opposed to what popular misconception makes them out to be. The misconception is caused by the scientists themselves in many cases, because they imply if not state outright that science by definition is a quest for truth; each scientific discovery, they imply, leads humankind progressively to a greater degree or amount of "truth," and in so doing diminishes the credibility of and need for religious belief, which they lump together with "superstition." That is a false paradigm, however. Science is actually leading us progressively only to a greater understanding of the physical, natural universe; it tells us nothing about whether there is anything beyond that. If scientists would simply acknowledge that religion, by definition, has no choice but to deal in the supernatural and metaphysical—whether speculating or pontificating dogmatically about a god or gods' role—they could then ignore it and not get drawn into debates about it in which there can be no winners.

So why do they keep falling into this trap, asks Singham? Because they foolishly try to defend what they think is THE truth. They think they are protecting the innocent victims (American school children) from the pseudoscientific quackery of charlatans. That is not their job, however. Nor is it wise, because these Genesis believers possess SOME truth, too, and it needs to be heard. The scientific community will shoot back that this "truth" lay only in the field of ethics and values, not in physical science, so they should NOT be heard on the issue of origins. Trying to silence the opposition is the wrong course of action, however. Science should be above that. Otherwise, scientists become the equivalent in reverse of the Church during the time of Copernicus and Galileo—a body devoted to suppressing dissent rather than proclaiming "truth." Scientists should merely keep "doing" science, and let Genesis believers keep trying to rebut their work. "But what about the schoolchildren???", they will ask in consternation. Singham answers in effect, so what if Genesis believers want to teach creationism as a viable alternative to what secular science tells us about origins? If it's not *really* true, students will eventually see through it and abandon it. No harm done. It basically just becomes an exercise in critical thinking, which is ultimately what the public school system should be cultivating anyway. I agree. I found Singham's case to be logical and practical from a scientific and a legal standpoint, as well as insightful from a psychological and sociological standpoint.

The first book I pulled from the shelf that dealt with the clash between science and religion offered help that I would need in countering some of secularist arguments: *Religion and Science: Conflict and Synthesis* (S. P. C. K., 1964), by I. T. Ramsey. It is a brief little collection of lectures by a professor at Oxford University which falls into the scholastic tradition of trying to reconcile science and religion. In order for me to write *The Explanation* or any book like it, I had to grapple with this most basic issue—can an ancient religious tradition such as Judeo-Christianity, which is set in scripture and thus cannot be easily altered, be harmonized with modern science which is constantly changing by growing in knowledge? Ramsey helps answer that question. As a theologian, he of course says yes. He takes the view that the two operate in different spheres but both spheres are equally necessary, and each complements the other. He disagrees with secular scientists who say that religion is nothing more than a superstitious belief system that people use to fill the void of scientific ignorance until such time that said ignorance can be replaced with scientific knowledge.

Ramsey offers dancing as an illustrative example of why this mode of thought is wrong. If dancing were to be viewed through the cold, unfeeling lens of physical science, all we would see is a human making a series of strange gyrations. What we would see would indeed accurately describe the phenomenon being witnessed, but it would tell us nothing about why the person was making those strange movements, and in fact it never could tell us why, no matter how hard the scientist tried. Religion, he says, puts feeling and emotion into the mix, and it is just as instinctive in humans to believe in God and to move under an invisible influence as it is for humans to feel compelled (and they don't know why) to dance. Sure, social scientists such as sociologists or anthropologists might be able to explain this behavior called "dancing" as a cultural and psychological phenomenon, but that is quite different than a physicist watching the motion of a dancer and trying to explain it through the cold logic of mathematical calculations! Secular scientists make the same mistake, implies Ramsey, in assuming they can account for the creation and operation of the universe without a creator and operator who supplies the reason for the phenomenon. Theologians and philosophers serve the same basic function as the sociologists and anthropologists in the analogy of the dance. I agree with this way of framing the issue, which is why I could muster the confidence to tackle this enormous and daunting subject of the origin of the universe and life by writing *The Explanation*.

Whereas the above-mentioned books are mostly tackling the religion versus science topic from a philosophical perspective, *The Spiritual Brain: A Neuroscientist's Case for the Existence of the Soul* (Harper One, 2007) by Mario Beauregard and Denyse O' Leary confronts it with medical evidence. This is a marvelous treatise that demolishes the "materialist" point of view in science and lends credibility to spiritual and/or metaphysical explanations of life. "Materialism" is basically another name for what once was commonly referred to as scientific "naturalism"—the belief that all things, regardless of how inexplicable at the present, ultimately have a physical or material explanation; i. e., there is no God, there is no spiritual plane, there is no life after death, etc. The authors distinguish between the "mind" and the three pounds of gelatinous gray mass called the human "brain." They show that we are more than biological robots programmed for survival by evolution, as materialist would have it. More importantly, they prove through a tremendous amount of evidence that the field of science is ruled by materialists who preach materialism as dogma and have closed their minds to any other possibility. The authors indict all the major scientific societies that dismiss paranormal and parapsychological phenomena as quackery even when all evidence suggests that some things cannot be explained through conventional materialism. They argue convincingly that the quest for truth has long since been replaced by determination to uphold materialism at all costs; no evidence that questions the assumption that the only rational explanation is a material explanation is permitted within the esteemed academy.

One example that Beauregard and O' Leary cite repeatedly of a non-material phenomenon that a) affects the physical world, b) is absolutely verifiable by all credible research, and c) thus flies in the face of materialism is the placebo effect. In innumerable cases, the placebo effect is just as important to the healing process as any medicine that a doctor might prescribe, yet it has no material explanation. It is simply the mind bringing about the healing of the body. It is, therefore, a perfect example of mind over matter—not brain over matter, for the brain is merely a tool that plays host to the mind. The mind is consciousness, and the authors make a strong case for human consciousness continuing after the brain has died. This book is no lightweight attempt to take on the big boys of standard secular science; it is a major, serious, damning indictment of those who automatically rule out the supernatural before even checking the evidence.

George Gallup, Jr., of Gallup Poll fame, gives statistics on how many Americans believe in life after death in *Adventures in Immortality* (McGraw Hill, 1982). He shows that Americans are far more believing than are people in any other industrialized nation. Within the U. S., the Deep South, which has the lowest education rates, has the highest rate of belief in the afterlife. Gallup spends most of the book reviewing cases of Near Death Experience (NDE). I confess that reading about those who have gone to "heaven" (or think they have) and returned to tell about it has comforted me over the years and bolstered my faith. Secular scientists try to debunk these NDE's, of course, saying they are hallucinations of a dying or oxygen-deprived brain. They cannot prove that, however, any more than those who say they have gone to heaven can prove their claims. While I can see how skeptics would think that some claimants are just repeating what they've heard other NDE'ers say, or spouting religious language from their own church background to describe their experience, it seems unlikely to me that every case of NDE is fabricated or just a hallucination. I think we can and should take the NDE phenomenon as a whole as evidence that when we die, our spirits (or consciousnesses) go *somewhere*.

Interestingly, Gallup notices that almost none of the people he polled who claimed a NDE had gone to "hell" or some such bad place. He points out that some Fundamentalist Christians take this as evidence that the ghostly figure in white whom many NDE'ers see is not Christ, as they assume, but Satan, who transforms himself into an angel of light in order to deceive them. They say this because almost no NDE'ers ever come back with instructions from this otherworldly being to give their lives to Christ or go spread the Gospel or some equally "Christian" thing. Instead, they seem to have a generic feel-good experience in which they are loved and validated and made to feel at peace pretty much regardless of what kind of life they have lived, who they have hurt, whether they were devout or not, etc.

Shawn Lawrence Otto's *Fool Me Twice: Fight the Assault on Science in America* (Rodale, 2011) discusses the politics of science in 21st Century America. It takes a leftist view which is critical of Evangelical/Fundamentalist Christians in the political arena. Of importance here is the author's take on how rock solid (pardon the pun) the secular belief in the age of the earth is. He claims that it "isn't something we believe, it's something we measure, like the distance between Minneapolis and Dallas" (27). It also gives the standard anti-Christian-nation version of American history, which is personally insulting to me in its ignorance and arrogance, considering that I wrote the book on the subject: *Christian Nation? The United States in Popular Perception and Historical Reality* (Praeger, 2010).

Stuart A. Kauffman's *Reinventing the Sacred: A New View of Science, Reason, and Religion* (Basic Books, 2008), is similar. It is anti-creationist in tone. The author claims that "almost all scientists" believe that life arose naturally and spontaneously in the universe, not by divine fiat. Moreover, he believes that scientists will "create life anew sometime in the next hundred years" (46). The first claim is demonstrably false; there are thousands of serious scientists in the world who believe that life originated with God, and there are thousands more who admit they just don't know, so they are agnostic. On the second claim, I simply disagree. Time will tell.

THE ATHEIST CASE AGAINST GOD AND CREATIONISM

Among the many atheist manifestos and anti-Christian books I looked at in researching *The Explanation*, the one that stands out the most is Bertrand Russell's *Why I Am Not a Christian: And Other Essays on Religion and Related Subjects*, edited by Paul Edwards (Touchstone Books, reprint 1957). Russell was a brilliant intellect and a man ahead of his time. He laid the foundation for a later generation of militant atheists who see it as their duty to oppose religion generally and Christianity specifically. In some ways, he was a wise man. He correctly identified some of the major problems extant in the church world, for example, circa the 1920s through 1950s. He points out hypocrisy, corruption, and dissension as things that have always abounded within the various sects and denominations, and wonders why if Christianity is true no one has yet figured out how to rid the religion of them. Russell's case against Christianity was so solid, in fact, that I can easily see how any person not firmly and completely grounded with roots growing deep would abandon the faith upon encountering it for the first time. His arguments countering the major Christian apologetics theories of the 1920s show how far ahead of the curve he was, for some of the same arguments have been revived by Richard Dawkins, Christopher Hitchens, and others in more recent years.

Despite his impressive array of knowledge and his unusually high degree of critical thought, Russell was not without flaws in his logic and not without blind spots in his formulation of philosophy. For instance, in his case against what we call "Intelligent Design" today (although it did not have a name in the 1920s), Russell asks rhetorically whether an intelligent designer couldn't have come up with a better world than one which sports the Ku Klux Klan and Fascists. (He didn't know, of course, when he made those remarks in 1927 that both were about to be relegated to the dustbin of history and replaced by atheistic Communism as the main purveyor of murder and destruction in the world.) That it never occurred to him how such misery-inducing political groups were products of a fallen, sin-filled world should not surprise us. Most likely he did consider it, but dismissed that notion as not worthy of even mentioning. In making his argument against Christianity, he explains how so-called Christian nations or governments have routinely chosen not to do what Jesus said to do—turn the other cheek, love your enemies, give to the poor, etc. It should likewise not surprise us that in formulating his argument he fails to distinguish between the rules by which nations will be held accountable by God and individual Christians will be held accountable. After all, some of the best Christian thinkers today cannot easily make that distinction, either, and perhaps most rank-and-file Christians cannot make it at all.

A final point to make about Russell's anti-theology (atheology?) comes in relation to his comments about Samuel Butler's *Erewhon* (1872) and *Erewhon Revisited* (1901). In these books, a man named Higgs discovers and gets stuck in a strange, backward, remote country, but eventually escapes from it in a hot air balloon, only to return twenty years later and find that a new religion has sprouted there based on his ascension into the heavens. Russell says, in so many words, that the Christian religion is based on something equally foolish. Yet, he, of course, could not have anticipated in 1927 that a whole cottage industry would soon emerge taking seriously the notion that visitors from space landed on earth in the past, were mistaken for gods, and had religions created in their honor. Nor could he have known that this cottage industry would eventually grow into an American pop culture phenomenon, and that the ancient alien theory would someday be used by me to explain how to understand Genesis accurately. Even so, it goes to show how an otherwise wise man can be utterly foolish in his thinking when he starts poking fun at those with whom he disagrees.

A book that is very much in the same vein as Russell's is one by the late Christopher Hitchens called *God Is NOT Great: How Religion Poisons Everything* (Twelve, 2009). To begin discussing it, there are several positive things to be said for this book and its author. Of the small number of militant atheist authors on the market who have made it big in recent years, Hitchens was the most likeable. He was not as snotty in his intellectualism as Richard Dawkins or the others. He had a humbleness about him that lent credibility to his views, which is one reason I couldn't put this book down. He concludes the book by saying he spent his whole life writing it. This is not hard to believe, considering how sweeping is the coverage of the world's great religions herein. Much of his work was done personally, traveling from nation to nation, culture to culture, and experiencing these religions first hand. His intellect was unquestionably in the top tier; there is no disputing his mental capacity for understanding the topic he addresses and for opining on it. Few people in the world, or the history of the world for that matter, would be able to match him in this regard. His vocabulary and command of the English language was off the charts. I learned a bunch of new words from reading this book. Some of them were academic-theological-philosophical terms that would only be applicable in a discussion among ivory tower types (like myself and Hitchens. . . and perhaps YOU, if you are actually reading this bibliographical essay!). But many of the words were just plain dictionary-English terms for everyday use, but which no one but the most educated and erudite actually use. I will begin using some of them as a result of reading this book.

Now, for the critique. Hitchens does the very thing in this book that he chides the "faith-based" people of the world for doing throughout history and now—being dogmatic in a belief system that cannot be proven. He rails against those who claim there is a God and will do just about anything to convince or coerce everyone else to believe it. Then he turns around and trashes everyone who is so "stupid" as to disagree with him that there is no God. This is the problem with militant atheists; they do the same thing in reverse that they are always complaining about! (It goes like this: "If you don't agree with me, there is something wrong with you, and you and your ilk are therefore responsible for all the evils in the world!")

Admittedly, the problem throughout most of human history has been just what Hitchens describes: the hyper-religious people have demanded that everyone agree with them or else. We live in an age, however, when the tide has turned and the militant atheists are now on the offensive, and they are thirsty for blood in return. One need look no further than the Communist regimes of the world from the 20th century to present which impose atheism on their subjects under penalty of death. Hitchens addresses this fact in a roundabout way, but does not do justice to the topic. He more or less excuses Soviet atrocities under Stalin, which, if they could be counted accurately, would undoubtedly outnumber most of the murders committed by the hyper religious throughout history around the world combined. The best estimate is that Stalin deliberately murdered more than 10 million (which is about 3 million more than Hitler, by the way). So forgive me if I'm not buying into Hitchens' argument that the world would be better off if there were no religion and everyone was just a supposedly enlightened atheist like he was.

The main argument against what Hitchens proposes here (that the world would be better off if we were all atheists) is that most people in the world do not have the mental capacity that he had. IF, and this is a huge IF... IF all people were as "smart" as Hitchens and Richard Dawkins and the intellectual elites who write these atheist manifestos, maybe just maybe the world would be able to progress under atheism. In the real world, however, where exactly one half of all the people in the world possess a mere double digit IQ, atheism produces chaos. Space and time fail me to fully explain this here and now, but you can figure it out on your own, because if you are reading this, then you are in the half of humanity that possesses a triple-digit I. Q.!

Hitchens' premise is actually wrong here, as evidenced by his subtitle. Throughout the book, he shows how the various big religions of the world have poisoned their own societies throughout history. But he does not distinguish between "good" religion and "bad" religion. He says all religion is in effect bad. The Christian riposte ("riposte" is one of those new words I learned from reading this book) is that James 1:27 says, "Pure and undefiled religion" is to do good deeds, such as take care of the widows and orphans, etc. Christians, therefore, are capable (in theory) of distinguishing between "good" and "bad" religion. Why can't militant atheists do the same? Answer: because it would undermine their argument against religion, or at least muddy up the waters in making the case against it.

The one person among all others discussed in this book about whom I know a great deal is Martin Luther King, Jr. Hitchens holds him up as a paragon of virtue to a greater extent than any other person in human history. Yes, you heard that right. Read the book for yourself if you don't believe me. Hitchens admits that MLK had some flaws (he was a notorious philanderer, and he plagiarized his doctoral dissertation—meaning he stole most of it), but they are all more than cancelled out by the great good he did not only for his own race but for all of humanity. Well, I wouldn't disagree with that characterization except that Hitchens goes on to say that, despite MLK's "Reverend" title and doctorate in theology, he was in no way a "Christian." Everything about his worldview and modus operandi, says Hitchens, was that of a secular humanist. I couldn't disagree more. In fact, I would go so far as to laugh in his face for that whopper! Again, space and time fail me to explain this.

I could drag this critique out, but instead let me conclude with one more observation. Hitchens sets up Socrates as another protagonist for atheists to look at as an exemplar. Socrates stood up against the hyper-religious powers that be and was put to death by them as a result. At least, that's how Hitchens portrays the story. But that's not exactly what happened. Socrates actually spent his time opposing the Sophists, who were the educated elites who (like Hitchens and the intellectual elites of modern times) believed they knew all the answers. Anyone who disagreed with them was a threat to their power. Then there's the part about Tyrants in the Athenian polity at that time. Religion played a small role in all this, but politics and power played the biggest part by far. The truth is, Socrates challenged the establishment of his day, which was partly atheist and partly polytheist. Socrates himself was an agnostic truth-seeker who leaned toward monotheism (at least that's what we would call it today). So holding up Socrates as an example of an atheist who was put to death by religious authorities is just twisting and spinning the facts to help make a case.

Jesus also opposed the power structure of his day and died for it. MLK, to some extent, did the same. The reality of the subject, therefore, is that it is not "religion" that has poisoned everything throughout history and into the present. Instead, it has been and still is power that has poisoned everything (see J. Rufus Fears, *The Wisdom of History*, lecture series, The Great Courses: http://www.thegreatcourses.

com/courses/wisdom-of-history.html). It has been said rightly that power corrupts, and absolute power corrupts absolutely (to paraphrase British historian Lord Acton). That is why when the Roman Catholic Church had a monopoly on power in Europe, it reached the height of its corruption in history. That is also why when Muslim Fundamentalists have a monopoly on power in certain countries today, they are now reaching the height of their corruption. And finally, it is why when southern racists held absolute power in the USA, they reached the height of their corruption. Bottom line: the safeguard against corruption on a grand scale (genocide, torture, and miscellaneous other human rights abuses) is not to get rid of religion, but rather to check the power of ANY group that manages to get control of a nation or region by having other equally powerful forces as deterrents. Do you really think that if the atheists got ALL of the power, they wouldn't do the same or worse crimes against humanity than were ever perpetrated in the name of Jesus or Muhammad? If you think not, I refer you back to the Soviet Union under one Joseph Stalin. I rest my case. See also Richard Dawkins, *The God Delusion* (Houghton-Mifflin Harcourt, 2006).

CREATIONISM VERSUS GODLESS EVOLUTION

Before I knew I was going to write a book on this subject, and therefore began any "official" research on it, I was already aware that there are several Creation Science and/or Intelligent Design organizations in operation today. They are generally (with notable exceptions) ministries which publish their own literature rather than submit it to academic presses or other non-religious peer-reviewed publishing houses. The C. S. and/or I. D. organizations I am aware of have all come into being in the last twenty or so years. The

Discovery Institute in Seattle is one. Its subsidiary, the Center for Science and Culture, is another. Creation Ministries out of Australia (and Atlanta) is a third. Carl Baugh and David Rives, among others, operate yet other similar ministries. I have deliberately omitted the self-published literature from these authors and organizations, not because it is not worthy of being included in my own estimation, but partly because it is scoffed at by secularists as nothing more than religious propaganda, and partly because I want to be as objective as possible, despite my own personal Christian faith. In order for *The Explanation* to have maximum credibility and for me to have utmost integrity, therefore, I have omitted such sources from this bibliography. I must admit, however, that I have learned a tremendous amount of what I know on the subject of Genesis and ontology from them.

Among the most noteworthy defenses of Intelligent Design that is not self-published but is instead put out by a scrupulously and expertly reviewed secular firm is Stephen C. Meyer's *Darwin's Doubts: The Explosive Origin of Animal Life and the Case for Intelligent Design* (Harper One, 2013). This book is, as I see it, magisterial in its scope. It covers the whole history of Darwinian evolution and the many neo-Darwinian attempts to explain the "Cambrian Explosion" (the seemingly countless new types of creatures that suddenly sprang into existence without a trace of earlier evolutionary evidence in most cases). He deflates the various naturalist theories (punctuated equilibrium, self-organization, evo-devo, neutral evolution, epigenetic inheritance, and natural genetic engineering) one-by-one, showing how they fail to account for origins. I won't say he "debunks" them, because there is much that is good about all such theories, but he shows their limitations and in fact their inability to answer the question of where new "body types" came from. He, of course, argues that the Intelligent Design theory best fits the available evidence, and more importantly points out that it is not a "religiously-based idea" (338), despite the popular misconception. If... and this is a big "If"... If secular scientists would look strictly at the evidence, with no regard for religion one way or the other, they would agree.

Although Meyer works for the Discovery Institute, he has a Ph. D. in the History of Science from Cambridge, and his research is impeccable by anyone's standard, as evidenced by the fact that some secular publishers are willing to put their name on his books. That is not to say that such publishers endorse his conclusion that a non-material "mind" is responsible for life on earth. That conclusion is of course dismissed out of hand by the secular scientific community and, by extension, academia and the mainstream media at large because they accuse it of surrendering to mystical or magical thinking. Meyer answers his critics by pointing out that, despite claims to the contrary, all naturalists and/or scientific materialists believe in invisible and unexplainable forces, with gravity being at the top of the list. Nobody knows what it is or why it works, but all evidence suggests there is something to this invisible, inexplicable force because its results are constantly observable. Likewise, although natural science cannot (as yet) explain how a non-material mind can transfer information into the physical universe, that information clearly was not the product of spontaneous generation and eons of evolution; it was already here when the Cambrian Explosion occurred.

Meyer shows that secular scientists admit their inability to explain the origin of new body types routinely in their own prestigious peer-reviewed journals, but somehow those admissions never make it into popular culture. Instead, the defenders of Darwinism continue defending it, and the mainstream media, the government, most public universities, most public school administrators, and most textbook publishers serve as their echo chamber. These defenders include the National Academy of Sciences, the American Association for the Advancement of Sciences, the National Association of Biology Teachers, and the Smithsonian Institution, among others. This defense-and-echo routine should not surprise us, though, since the same thing happened throughout the English-speaking world immediately upon publication of Darwin's *Origin of Species* in the mid-1800s, as Meyer shows. None other than the greatest paleontologist of the 19th Century, Louis Agassiz of Harvard University, pointed out this Cambrian explosion mystery way back then, only to be drowned out by voices of his peers who were perfectly willing to overlook this defect in Darwin's theory since there were so many other compelling and correct things recommending his overall argument about natural selection.

Meyer's book is replete with interesting information and common sense observations and insights. To paraphrase a few examples: 1) the fact that the term "evolution" is used much too loosely and generically in popular culture, and that is one main cause of all the controversy about it; if people would be precise and differentiate between "micro" and "macro" evolution, a good number of arguments could be avoided. 2) Darwinism can only explain the survival, not the arrival, of the fittest; the question of what started it all lies beyond the knowledge of secular science. 3) Darwin himself admitted that he could not account for the Cambrian Explosion, but it has never given pause to his defenders much less stopped them in their tracks. 4) All evidence suggests that new species could not have come into existence through "big" mutations, because all big mutations cause deformity (and generally sterility or other traits that make reproduction less likely) in a species; therefore, the question of where all those new life forms came from in the Cambrian Explosion is every bit the mystery today as it was in Darwin's time. 5) Darwin's observations of microevolution in various species on the Galapagos Islands does not prove macroevolution, because in the end a finch is still a finch, regardless of its beak shape, and a tortoise is still a tortoise, regardless of its neck length, etc.

One of the main points of Meyer's book is to show how scientific materialism disallows paleontologists and biologists from invoking mystical, magical, invisible forces to explain the origins of life. Interestingly, Charles Seife's *Alpha and Omega: The Search for the Beginning and End of the Universe* (Viking, 2003), a book not at all designed to aid the Intelligent Design cause, shows that astrophysicists do that all the time. When they can't explain something about the character of the universe, they invent something to explain it. From ether in the 1600s to dark matter in the 1900s, cosmologists tend to agree that something invisible and undetectable makes up two-thirds of the universe. Since they can't see it, they look at its effects on the one-third they *can* see. Seife puts it like this: "As distasteful as it might be to believe in something that is invisible and, so far, undetectable, it is the best alternative. . . . Astrophysicists are forced to accept the existence of dark matter. They can only explain their observations of the universe if they invoke its presence" (101-102). Wow! Stop and let that sink in for a moment. They can only explain what they observe in the physical realm by invoking an invisible, undetectable presence. How is that so different from believers saying an invisible, undetectable being called God is behind creation?

Outside of those stunning statements, Seife's book is a rather standard scientific account of the history and current status of cosmology. It begins with the tired, old bashing of the medieval Catholic Church which stood opposed to secular astronomy. Then Seife takes the position (or so it seems) that the eventual triumph of science over the church's Aristotelian view of the heavens has proved the Bible to be nothing more than ancient fictional literature. He quotes Psalm 148 which speaks of "waters above the heavens" and the story of the sun standing still in Joshua Chapter 10 as evidence. Upon closing the case against the credibility of the Bible, he moves on to discuss the usual materialistic theories of the origins and fate of the universe.

Before any of the current C. S./I. D. organizations existed, there were others which I had never heard of prior to picking up Frank Lewis Marsh's *Life, Man, and Time* (Outdoor Pictures, 1967). The Evolution Protest Movement out of England was one. The Creation Research Society out of California was another, and a third was The Bible-Science Association, Inc., out of Idaho. How many others there may have been over the decades, I don't know. Marsh must have been a leading name in the field in the 1960s, however, based on his book. He stated many of the arguments that later creationists have expanded upon to show the shortcomings of the theory of evolution. Marsh lamented the Darwin Centennial Celebration that occurred in Chicago in 1959, quoting one of its sponsors who seemed to worship Darwinian evolution with just as much praise as any evangelic Christian has ever worshiped God. During those decades when Marsh was writing and teaching—prior to the emergence of our current wave of creationist organizations, when Darwinists scarcely had any serious opposition—it must have been harder to be a creation scientist than it is today. Marsh strikes me as having been a lone voice crying in the wilderness, and a simple Google search of his name reveals that to have been the case. He was certainly one of the few in that day.

Some of the most recent, cutting edge, secular scholarship on this subject has been done by Sean B. Carroll. He believes that creationism is all bunk, and he is glad to debunk it, but that isn't the primary purpose of his scholarship. His book *The Making of the Fittest: DNA and the Ultimate Forensic Record of Evolution* (W. W. Norton & Company, 2006) taught me (although I'm sure I must have learned this years ago in a science class) that there are actually three different "domains" (most basic forms of life) on earth: Eukarya (basically plants and animals), Bacteria (what we think of as "normal" microscopic life), and Archaea (microscopic life that exists in such extreme conditions, such as boiling water, that it does not have much in common with any other form of life). Within these three domains are five "kingdoms" (slightly more specialized forms of life): plants, animals, protists (single-cell organisms), fungi, and bacteria. Sometimes the lines separating these various life forms are blurry. Ironically, as scientists discover new varieties and try to fit them into ever more specialized categories, the lines do not come into sharper focus but rather grow increasingly obscure. See also the whole catalogue of book written by the late, prolific, secular evolutionist Stephen J. Gould, but perhaps start with *The Structure of Evolutionary Theory* (Belknap Press, 2002).

JEWISH INTERPRETATIONS OF GENESIS

Joel M. Hoffman, *And God Said: How Translations Conceal the Bible's Original Meaning* (St. Martin's Press, 2010) makes hay of the many problems inherent in translating the Bible from one language to another. Many of the specific examples he provides of well-known mistranslations come, not surprisingly, from the King James Bible. Most of them never confused me or had me believing some false doctrine merely because of a faulty translation. I think I speak for the vast majority of Christians like myself who were reared in Bible-reading, Bible-believing churches and Sunday School classes when I say these examples have never caused them to misunderstand the real meaning of the passages in question, either. Take the 23rd Psalm phrase "I shall not want." Hoffman makes much ado about how it means "I shall not lack for anything." Yet, it seems to me that everyone who grew up reading the King James Version or hearing it preached from understands that. Only a novice picking up the Bible for the first time might be confused by that. Or a reader for whom English is a second language might have difficulty with it, but it doesn't really seem like a big deal to me.

Other points that Hoffman makes are a big deal, however. He cites the great secular scientist Carl Sagan as an authority on how to interpret the Genesis passage about Eve being "punished" by God through pain in childbirth. Sagan noted that humans are the species that seems to have the most pain in childbirth, and the reason is the disproportionately large head of the human baby passing through a small birth canal. He ties this in with the fact that Eve ate from the "tree of knowledge" (not from the "tree of the knowledge of good and evil"). My disagreement with Sagan and Hoffman on this point is first that God did not actually *punish* Eve but rather informed her of the *consequences* of her action after the fact, and second that the "good and evil" part is the key, because it pertains to sex.

Judaism (George Brazillier, 1962), edited by Arthur Hertzberg, contains some interesting thoughts from the Jewish perspective on the nature of God. First, Hertzberg notes that "The Bible does not regard God's existence as something to be proved" (46). That is a statement which is profound in its simplicity. To put it another way, the Bible was not written by or for philosophers, which as a category of scholars tend to speculate about God's existence and/or nature and thus tend to elicit skepticism in their readers. It was written instead by theologians—not the kind that we think of today who are trained in seminaries and Bible colleges, but rather the kind that start with belief in God and with childlike faith and work outward from there. Theologians take it for granted that there is a God, and they build their philosophy or religion off of that assumption, trying to instill faith in their readers. Jewish scholars come in both categories; some are philosophical, speculative, and skeptical in their approach, and others are theological, simple, and faithful.

Hertzberg offers some long quotes from medieval Jewish scholars Saadia, Maimonides, and Hayyim ibn Musa that illustrate this variety of thought. He quotes Musa, who lived in Spain during the time of the Spanish Inquisition, as one who saw the danger in presenting philosophy in place of theology in sermons given to regular folks who go to synagogue to have their faith bolstered and their spirits refreshed. Musa once witnessed a rabbi preaching "in a speculating manner—in the manner of philosophers" to the aggravation of many of the regular folks in the congregation. One finally stood up and interrupted the sermon, declaring that he had been beaten and wounded by the Christian authorities of Spain in a pogrom in Seville merely for quoting the Torah: "Hear, O Israel, the Lord our God, the Lord is One." Having suffered greatly for his faith, he did not appreciate a preacher who wasn't as sure of God's existence and as simple in understanding God's nature as he was. He chided, "I do not want to go on listening to this sermon." With that, he walked out, and most of the other congregants followed him.

This passage illustrates the danger I potentially face in publishing *The Explanation*. There are many, like the man in this story, who don't appreciate speculative writings like *The Explanation*. They prefer for preachers and theologians to keep it simple and make it impossible to misunderstand. While I respect that mode of thought and am sympathetic to it up to a point, I have chosen to throw caution to the wind and publish this philosophical, speculative treatise anyway. I have done so because I believe in critical thinking. As a college professor who teaches History and related Humanities and Social Sciences courses, I believe it is important to push students to think harder and deeper about whatever subject we happen to be studying at any given time than they would otherwise naturally do. Human nature is to choose the path of least resistance, to gravitate to the easiest way to get from point A to point B in life. I have found from years of teaching that students will not learn any more than they are required to know, except in rare cases where a particular one has a special interest in the topic. They have to be forced to learn what they "don't know that they don't know," in other words. In a sense, I sympathize with them because, after all, once any of us learns to read, write, and do basic math, the rest is optional. Having learned reading, writing, and arithmetic in grade school, college students are choosing to learn more out of their own free will. And that's where I come in with critical thinking.

In preaching and/or writing books of theology from which others might preach or teach, there is a need both for simple words of faith or inspiration and for works of erudite scholarship that stretch the bounds of our thinking. The one is akin to learning and then applying elementary school reading, writing, and arithmetic skills; the other is like going to college and learning to plumb the depths of whatever subject is under consideration in any given course. *The Explanation* is a college-level critical thinking piece. It is not for everybody. It is not for folks who are happy being simple, uncritical thinkers who believe what they believe and don't want to be confused with facts that might contradict their beliefs. It is not for the spiritually uneducated or undereducated who choose to remain uneducated or undereducated. It is for those who choose to push the boundaries of their thinking further and further.

One reason I believe this is important is because everyone around the world is born into some religious culture, be it one of the many types of Protestant Christianity, or of Catholicism, Judaism, Islam, or whatever. In America currently it is almost as likely that one would be born into a culture of agnosticism or atheism (which is itself a dogmatic belief system like a religion). It is natural, and it is foolish, to accept whatever religion one is born into without examining it carefully for defects once we are old enough to be able to think for ourselves. Here I will tie this essay back together with the Hertzberg book *Judaism* by passing along a quote he took from Saadia, which I have abridged with several ellipses thusly:

> There are two sorts of persons who believe in God. The one believes because his faith has been handed down to him by his fathers . . . . The other has arrived at faith by dint of searching thought. . . . the first has the advantage that his faith cannot be shaken, no matter how many objections are raised to it . . . . But there is a flaw in it . . . it has been learned without thought or reasoning. The advantage of the second man is that he has reached faith through . . . much searching and thinking. But his faith too has a flaw: it is easy to shake it by offering contrary evidence. But he who combines both kinds of faith is invulnerable.

This quote basically sums up my own thoughts and reflects my own spiritual journey. Having received the faith by being reared a Bible Belt Baptist, I had to examine my religion for defects upon reaching adulthood to decide if it was really worth believing, or to see if it was, as many who discard it have decided, no different than believing in Santa Claus or the Easter Bunny. If after allowing the intellectual forces of secular science, psychology, history, and such free reign to poke holes in my bucket of beliefs, I have found that my bucket still holds water, then I am persuaded by both faith and reason that it is legitimate; it is *not* merely the equivalent of believing in Santa Claus and the Easter Bunny. This is essentially a medieval scholastic approach (think Thomas Aquinas), and it is in my opinion, the best (if not the only right) approach for intelligent adult American Christians today. I am writing *The Explanation* accordingly.

Rabbi Neil Gillman, *The Jewish Approach to God: A Brief Introduction for Christians* (Jewish Lights Publishing,

2003) discusses what is for Jews their most commonly quoted scripture—Deuteronomy 6:4. One popular translation says, "Hear O Israel: the Lord our God, the Lord is one," which is used in some versions of the Torah as well as some Christian Bibles. Gillman picks the verse apart, looking at its various possible English translations and trying to decide what it is really saying. He concludes that the conventional wisdom is correct; it means the God of Israel is the only God there is. In other words, the verse is a caution against polytheism, which makes perfect sense because that was the dominant type of religion in the world in ancient times and thus the main enemy of Judaism when the verse was written. To modern Christians, it was clearly not meant to be a prohibition against believing in the Trinity but rather in believing that the so-called gods of other nations were gods at all. Yet, the way Gillman frames the discussion, the emphasis is not on whether the so-called gods of other nations were really gods, but that regardless of whether they were or not, Israel was not supposed to worship them. The emphasis, therefore, is on Israel's having only one "god"—the Lord (YHVH).

Why is this important? Because it supports the theory that there were indeed extraterrestrial beings of some kind visiting earth, landing in various places, claiming to be gods, and either requiring or allowing worship from the local people. To put it another way, the Hebrew scriptures never said there were no other actual "gods" for Israel to choose from, but rather they imply that if Israel chose one of the other gods, they would be getting an inferior god, because YHVH was greater than them all. He was/is the God of gods, and the Lord of lords. Gillman does not say all this, but his discussion of how to interpret Deuteronomy 6:4 led me to make this connection.

Another point of fact I learned from Gillman is that the character we call "Satan" in English is actually called "ha-satan" (the satan) in the Hebrew book of Job. It is widely believed that Job is the oldest book of the Bible. If so, it would be the first place where this evil character is mentioned, despite the fact that his appearance in the Garden of Eden in Genesis indicates that he was at work long before Job was born. I found more on that topic in F. E. Peters, *The Monotheists: Jews, Christians, and Muslims in Conflict and Competition, Volume I: The Peoples of God* and *Volume II: The Words and Will of God* (Princeton University Press, 2003). It compares and contrasts the features of the three big monotheistic religions. I must say that if I were stranded on a desert island and could have only one book on the history of religion, it would be Volume II of this pair. It pretty much lays out the whole history and theology of the big three religions in a well-organized, easy to follow 400 pages.

As far as the creation story goes, Peters assumes the reader is familiar with the basics of the Jewish/Christian biblical account and focuses instead on the Muslim account in the Quran. He shows that the Quran says that an angel named Iblis refused to follow God's instruction of bowing before the man Adam He had created. Thereupon, God expelled him from heaven, along with the other angels who followed his lead. Now on earth, he becomes a "shaytan," which is roughly (but not exactly) the Arabic equivalent of our English Satan. Later in the text, however, Peters explains that our English word "Satan" should actually be rendered as the common noun "satan" (lower case s), because in Hebrew it means simply *an* adversary or *an* accuser, not necessarily *the* great or *the* ultimate personification of evil that we think of Satan as today. Even so, the conventional wisdom says that it was this particular adversary (enemy of God and man?) who showed up in Genesis and convinced Adam and Eve to eat the forbidden fruit. So he was a tempter and an instigator as well. Whether "the" adversary as Gillman puts it, or just "an" adversary as Peters puts it, I have portrayed him as the one responsible for the fall of humans into original sin, just as the typical, modern Christian view would have it.

On a different subject, the author states for a fact that "there were very few atheists in the ancient or, indeed, the pre-modern world" (4). This statement seems to reiterate conventional wisdom. However, it leads me to question whether we can know for sure how many atheists there were back then, or to be certain that the percentage of the population professing atheism was so low as to constitute "very few."

A third point of fact that I learned from this set is that the names we use in English for the books of the Old Testament, such as "Genesis," are actually Greek words that were chosen for use in the Septuagint (the famous Alexandrian third century B. C. translation) because they were descriptive of the contents. Hebrew scripture never used those terms, and Hebrew readers always refer to each book by whatever the first line of the book says. Our title *Genesis*, therefore, is actually *Bereshit* in Hebrew, which simply means "In the beginning."

A fourth point raised herein is the various appellations used for the creator-God throughout the Old and New Testaments. Peters offers that YHWH ("Lord") was used first, then the plural proper noun Elohim ("God"), and then Adonai ("My Lord"). The use of so many different monikers for the monotheistic deity in the Hebrew scriptures led early Christian writers to have to fit these variations into their theology. The common understanding in the Church came to be that the pre-carnate Jesus was YHWH, the creator, which comports with John 1:1, whereas Elohim is seen as the spirit-Father of YHWH, and Adonai would be reserved for certain special occasions but would refer to YHWH. This theology is rife with all kinds of brain-twisting complexity, of course, but that is the very type of thing *The Explanation* is designed to untangle.

A notable treatise on that topic is Harold Bloom's *Jesus and Yahweh: The Names Divine* (Riverhead Books, 2005). Bloom is a venerable old literary critic with a long list of books to his credit, more than one of which I've read. As a Jew, he readily admits that he doesn't believe the Gospel, but he finds the Bible a fascinating collection of ancient literature. As a literary critic, he picks the Bible apart, as he would do any other book or series. He finds the Old Testament to be much superior to the New Testament as a work of literature, although he prefers the Jewish *Tanakh* version of it to the Christian canonical version. Within the New Testament, however, he sees the Gospel of Mark as the most historically credible of books and the Gospel of John as the least. He despises the writings of Paul and has no respect for this apostle who penned a third of the New Testament, considering him a *Tanakh* scripture-twister of the most egregious sort. More importantly, Bloom does not see Jesus and Yahweh as being at all compatible as literary figures. He points out in great detail all the things that seem wrong to him about the character of the God of the Old Testament and the God of the New.

This is, of course, precisely what one would expect from a non-believing ivory tower scholar like Bloom. The thought which kept recurring to me as I read this book was that this may be what the prophet Isaiah was referring to (which Jesus referred back to) when he spoke of a stumbling stone and a rock of offense. The God of the Old Testament does indeed seem quite different than the God of the New by any rational, objective measure—so much so in fact as to trip up all but the most die-hard Christians. I confess that this has been one of the biggest mental hurdles I have personally had to jump in becoming an apologist for the faith and the author of *The Explanation*. The way I have made that leap is by considering what I just said above—that scattered throughout the Old Testament are prophecies foretelling the coming of a new covenant that God would make with all mankind and which He would use the Jewish religion and Jewish messiah to deliver.

UNORTHODOX ONTOLOGICAL ARGUMENTS

Tying Jewish interpretations of Genesis to the next category of books is Gregg Braden's *The God Code: The Secret of Our Past, the Promise of Our Future* (Hay House, Inc., 2004). It makes a really interesting case for who and/or what God is based on the Kabbalah, the Hebrew alphabet, and gematria. The case has several steps which make perfect sense when following them the way Braden lays them out. It would be hard to do justice to it in a synopsis, but I will try. The Kabbalah is composed of three ancient books, of which the *Sepher Yetzarah* is said to be most important. It is certainly the one most germane to *The Explanation*. It is full of mysterious information that must be decoded by experts to be understood. When done properly, it yields clues that can unlock the secrets of the cosmos, creation, heaven, immortality, and the nature of God, among other things. It also corroborates and is corroborated by modern science, as Braden shows.

For laypeople to grasp the decoding process, we must know some basics, such as that the Hebrew language does not contain vowels, only consonants. Hebrew words have to be inferred from the context in which they are used. One Hebrew name for God in the Torah is written "YHVH," which translated gives us both "Jehovah" and "Yahweh" as possible ways to say and write God's name in English. This word, however spelled or spoken in English, means "The Lord." In Hebrew, each of the 22 letters in the alphabet has a numerical equivalent, which is where gematria comes in. Gematria is a type of numerology that Kabbalahists use to find patterns and associations between Hebrew words, which then can be used to decode secret messages embedded in the Torah.

By knowing certain secular scientific facts apart from any theological, philosophical, or numerological studies, Kabbalahists can fashion their overall explanation of all that exists. One of these facts is that DNA (which all living things have) is composed of four things: adenine, thymine, guanine, and cytosine, which are called the DNA bases ATGC. Each of these bases is in turn composed of four chemical elements: hydrogen, nitrogen, oxygen, and carbon. The Hebrew letters for these four elements are YHVG, which is almost the same as the name for God YHVH. The difference is one letter/element. In God's name, the letter for carbon is replaced with a second nitrogen. God's name, therefore, is composed of four invisible, odorless, tasteless gases, and according to Braden, God is himself composed of these four gases and has no physical components at all. This comports with the Christian theology of God being a spirit. What religious language calls spirit, in other words, scientific language calls gas or a gaseous substance. It also is understood by Kabbalahists as breath, wind, and/or air.

Now, to get to the crux of Braden's explanation, whereas all of the elements in God's name are gases, 3 of the 4 elements in all living creatures are gases while 1 is a solid. The only difference between God and all physical beings, therefore, is just one element. God is a spirit, but humans and all other living, terrestrial things, are solid. We are carbon-based, in other words, yet we are three parts God! In case I have not made Baden's explanation clear thus far, let me add that the Hebrew word for man or human is that which we translate into English as "Adam." The Hebrew letters for Adam are YHVG, which is the same as the four DNA bases!

There is much more supposedly secret knowledge that can be gleaned from Kabbalahism, but for my purposes in *The Explanation*, this will suffice. Whether any of this is true, real, and scientifically and theologically sound is a matter of opinion. Fundamentalist Christians will likely conclude that it is just cunningly devised pseudoscience and pseudotheology. I tend to think there is some validity to it. How much, I'm not sure. I have taken it into consideration, along with the myriad of other pieces to the puzzle, in arriving at *The Explanation*. See also Daniel Lapin's, *Buried Treasure: Hidden Wisdom from the Hebrew Language* (Multnomah Publishers, 2001).

Paul Dehn Carleton's *Concepts: A ProtoTheist Quest for Science-Minded Skeptics of Catholic, and Other Christian, Jewish & Muslim Backgrounds* (Carleton House, 2004) argues that belief in God is merely a human creation, stemming from our instinctive "Life Urge." According to this explanation, early humans created God and/or gods because they needed a parent figure and a way to make sense of an otherwise senseless world. So they projected their inner needs outward, using their imagination. Because the will to survive is the most basic instinct in all animals—including humans—it was a natural evolution for them to create an imaginary immortality for themselves in which their spirits or souls would continue to exist even after their mortal bodies died.

Frank J. Tipler's *The Physics of Immortality: Modern Cosmology, God, and the Resurrection of the Dead* (Doubleday, 1994) sets forth one of the most provocative theses I have ever come across--that God is and/or will be the culmination of all intelligent life in the universe at the end of time. Tipler calls this culmination "the Omega Point." The argument goes like this: the universe and space-time will end in about 100 billion years in what mainstream physics calls a singularity, which will essentially be a point of infinite density just as existed before the Big Bang. According to mainstream physics, this singularity will destroy all life in the universe. Tipler argues, however, that intelligence will have evolved and grown by then to the extent that it will control all of space-time and matter, such that it can alter the way the singularity occurs, allowing itself to survive and thus preserve in itself all information ever accumulated throughout existence for time immemorial. The Omega Point will be basically the equivalent of a computer database containing the essence or emulation of every living soul and all the knowledge, feelings, and memories each experienced. It can, in other words, resurrect (re-create) each human from the beginning of the species to the end of time, and, according to Tipler, it will. This is how immortality will be achieved, fulfilling the Christian prophecies of the resurrection of the dead.

Moreover—and this gets deep—light travels faster than any material thing in the universe, but information or intelligence is not material and thus travels at the speed of light. Since the Omega Point is the culmination of all information and/or intelligence, it has long ago (by our earthbound, mortal understanding of time) received the data of everything that has ever happened or will happen. It can predict what we think of as the future, because it *knows* it just as you and I can know the past. It actually *experiences* what we perceive as linear time—past, present, and future—all at once. That is how it (God) is omnipresent, omniscient, and omnipotent from our vantage point as humans on earth. We see God that way, and for some believers, we see ourselves as, in a manner of speaking, little pieces of God, which, according to the Omega Point theory, we are. Tipler's thesis is an example of how some scholars can envision God as being within the universe rather than above and beyond it and yet still be "God" in the Christian sense of the word.

I found out later in my research from reading *Heaven: A History,* by Colleen McDanell and Bernhard Lang (Yale University Press, 1988) that what Tipler is describing is basically just a scientific explanation of how "Process Theology" could work. We might say that one version of Process Theology was first developed by Joseph Smith, the founder of the Church of Jesus Christ of Latter-Day Saints in the early 1800s, although the term "Process Theology" was certainly not coined until the 20th century. Smith taught, and the Mormons as a whole teach, that individuals continue to learn and evolve spiritually and intellectually in their lives after death, and through this process eventually becomes gods. A more mainstream version of Process Theology—at least among seminarians and other academics—was developed in the early 20th century by philosophers and theologians such as Alfred North Whitehead and Charles Hartshorne. Their belief was basically that individuals in eternity are nothing more than the sum of a series of events or occasions that they experienced in life. The entity that records or remembers this series of events or occasions is called God. Ultimately, God *is*, as McDanell and Lang put it, the "Cosmic Consciousness into which every past event is incorporated" (348). While I can certainly see how this argument can be made logically and intellectually, and thus how we should take it seriously, I reject it in *The Explanation*. My thesis is that God must be and always have been supernatural and metaphysical, not a product of the natural evolution of intelligent life existing within an uncreated physical, natural universe.

Jim Holt, *Why Does the World Exist? An Existential Detective Story* (Liveright Publishing Corporation, 2012) is one of the deeper philosophical works I have ever read. It basically is a study on why there is something rather than nothing. Nothingness is seen as the opposite of what we observe in the physical universe, which is "something" (matter). The question is, why did something happen? Where did something come from? What came before something? Nothingness? It is really the ultimate riddle for philosophers and physicists alike, and there is no sure answer. The point is, without believing in a God that exists in a spirit realm, completely separate from the physical universe, there is no way to conceive of or imagine pre-Big Bang existence.

Leon Lederman's *The God Particle: If the Universe Is the Answer, What Is the Question?* (Houghton Mifflin, 1993) is among the most reader-friendly works of profound scholarship [normally "reader-friendly" and "profound scholarship" are oxymorons] ever written. A witty yet erudite history of quantum physics up to the early 1990s, it shows the as-yet unsolved mysteries of the subatomic universe. It also shows that as physicists and chemists dig ever deeper into the atom and study the *things* that comprise it—whether particles, waves, or otherwise—there is *something* holding all the parts together and making them move that they cannot understand. They will never be able to "see" the inner workings and interaction of these parts, because any instrument from our human-size world that were to be inserted into the quantum-size world would destroy the very thing it was designed to study. Scientists therefore must infer from evidence other than the visual kind what goes on inside the atom, what its parts "look" like, etc. What they do know is that, if it were possible somehow to have instruments to see that far down into the quantum world, there would be vastly more empty space than matter. What is the force that keeps those subatomic *things* in perfect balance, attracting their attractive mates, repelling their repulsive counterparts, and maintaining the exact right amount of empty space to make the various atoms, which in turn make the various molecules, which in turn make the world we are able to see?

I opine that God created all that exists in the physical universe, from the largest galaxies to the smallest quarks, so size is irrelevant to him. Or better put, it is not a limitation to him like it is to us. He can get down to the size of subatomic things. He either established some natural law to govern the quantum world, or else some part or manifestation of him resides within that supposed empty space and holds it all together and makes it all go.

WHO WROTE GENESIS?

Isaac Asimov's *In the Beginning* . . . (Crown Publishers, Inc., 1981), looks at the creation story through the lens of the so-called Wellhausen school of theological scholarship. It breaks the story into two parts based on who this school of theologians says wrote the ancient Hebrew manuscripts that over time came to be codified as the orthodox creation story found in Christian Bibles. It says there were actually two separate creation stories written by different authors. One is called the "P-document" (short for Priest document), and one called the "J-document" (short for Jehovah and/or Judah document). The P-document, which was written later (the 500s B. C.), covers all of chapter 1 and the first three verses of chapter 2 in the modern book of Genesis, and the J-document, which was written earlier (the 700's B. C.) takes up the story there, which is why there seems to be a retelling of the creation story in different words and emphases in chapter 2.

Karen Armstrong's *The Case for God* (Alfred A. Knopf, 2009) takes the same approach but brings the theory up to date. It says that there were two authors living in Israel in the 700s B. C., "J" and "E" (short for Elohim), whose work was later combined into one, revised, and enlarged by other authors and editors called redactors, reformers, and Deuteronomists (who wrote the "D-document"). Then in the 500s B. C., during the Babylonian captivity, still other authors and editors called priests added even more and revised even more. This was the origin of the P-document, and it is this version of the creation story that we read in Genesis chapter 1.

For more information along these same lines, see *The Interpreter's Bible, Volume I*, edited by George Arthur Buttrick, et al. (Abingdon-Cokesbury Press, 1952). Although older than the Asimov and Armstrong books, it is more detailed than either on this topic. The theological school that these books represent is mainstream today among seminaries with a liberal (non-Fundamentalist) bent. This view basically says that we should not take the book of Genesis literally as if it were written by the finger of God. We should instead read it for what it is—mythology, plain and simple, which is what the best available archaeological and historical evidence tells us it is. According to this view, this mythology can be useful as a religious tool through which to teach life lessons, but it should not be seen as actual history much less as a scientific explanation for the origin of the cosmos or life on earth. Obviously, I do not subscribe to this school of thought. If I did, writing *The Explanation* would be an exercise in futility. On the same topic, see also Richard Elliott Friedman's *Who Wrote the Bible?* (Summit, 1987).

In opposition to this Wellhausen school is William Henry Green's *The Unity of the Book of Genesis* (Charles Scribner's Sons, 1910), which shows that as early as 1895 some scholars were doubting and disputing that the P, J, E, and D theory was true. Green argues that Genesis does not have different authors, but that the one author is simply telling three different stories: the generations (or history) of the heavens and the earth, the generations (or history) of Adam to Noah, and the generations (or history) of Noah to where the book ends with Jacob's descendents. Where Genesis ends, I see now, makes logical sense, because basically the narrator stops after setting up the conditions in which the political *nation* of Israel will be born in Exodus (which is separate from the *genetic line* of Israel which originates in Genesis). Green also claimed, contrary to what the mainstream view today is (if I understand it and him correctly) that "Elohim" is the proper name for God in the creation story, not YHWH. He points out that YHWH is the God of Israel, but Elohim is the God of the Universe. I tend to agree with that point of view, and it is one profound point on which *The Explanation* hinges. It seems important to me to note that I had already developed that point of view, or something close to it, independently before picking up Green's book.

Tremper Longman III's *How to Read Genesis* (InterVarsity Press, 2005) falls somewhere in between the two extremes above. First of all, it makes an excellent starting point for understanding Genesis. It tackles all the hard subjects, such as who wrote Genesis and whether it should be read as history or myth. It doesn't go into great detail in providing insights on these subjects, but it does give helpful information in overview. On the question of who wrote Genesis, Longman says Moses was probably the original writer but that later editors surely tinkered with it. One example he cites is where the narrator lists Abraham's home town as "Ur of the Chaldeans": the term "Chaldean" was not in use until several hundred years after Moses lived. He believes later editors added that point to clarify for Jewish readers of the day where exactly their forefathers came from. This makes about as much sense to me as anything I've read on the subject of Genesis's authorship. If Genesis is, as I believe, divinely inspired, that means God delivered a central message in it, and He gave that message to someone such as Moses, who then wrote it down and passed it along for others to tweak. When those others tweaked it, however, they did not change anything to do with the central message; they merely added points here or there for clarification of things within the original text that they thought other readers would be confused by.

SATAN

Many people who claim to believe in a good and loving God who takes us to Heaven when we die find it hard to believe in a purely evil character known as Satan or that God would send anyone to (or allow anyone to spend eternity in) Hell. 99% of psychiatrists believe in neither God nor the devil. M. Scott Peck, M. D., is one of the exceptions. This is fortunate for us because he was also a famous, highly-acclaimed author who, just before his death in 2005, went out of his way to provide evidence of the existence of Satan in *Glimpses of the Devil: A Psychiatrist's Personal Accounts of Possession, Exorcism, and Redemption* (Free Press, 2005). He says that Satan, which he calls an "it" rather than a "he," is like God in the sense that both are spirits, not material beings. To manifest in the physical world, Satan requires that somebody (or some body) serve as host. Preferably it would be a human body, but in times when one is not available, Satan or any of his minions can inhabit an animal. In order to inhabit a human body, Satan must be invited in or allowed in by the host; it cannot take over a body by force. It can be cast out, but only by the most determined and experienced exorcists.

Throughout the book, Peck offers insights that I take as pearls of wisdom. One that hit me where it hurts as I was writing *The Explanation* is that "Human beings were not created to have all the answers" (52). Mystery represents truth, he says. Demons often try to bring confusion into the minds of humans by pretending to give them all the answers. Often they do this in the guise of science. This gave me pause briefly about what I was trying to do in writing *The Explanation*. However, I overcame the reticence by weighing in my mind the pros and cons of publishing this book and deciding the former outweighed the latter. If I were pretending to *know* and proclaiming this supposed knowledge as truth dogmatically, I would be guilty of what Peck cautions against; I would be guilty of deception. If instead I offered *The Explanation* as a plausible alternative to remaining in ignorance or having theories that come only in pieces and parts that do not connect together, I would be doing the world a service by saying what *might* be the truth, not trying to deceive it.

In one of Peck's attempted exorcisms, he notes that the possessed victim exhibited all the characteristics of a serpent. He wondered "whether this phenomenon might somehow be representative of the snake in the Garden of Eden—of the devil in its earliest recorded appearance" (172). This is important to my thesis because it shows that an imminent psychiatrist who was a trained, scholarly, scientific thinking medical profession took the Garden of Eden story seriously and perhaps even literally.

Peck speculates that free will is the thing that causes human beings to have the image (but I say instead the "likeness") of God, and that is the reason Satan hates us so much—because angels were never given that free will, and that makes humans higher in the hierarchy of creation than they. He admits that this speculation is contrary to Christian orthodoxy but notes that it is essentially what Islam teaches. This speculative theology is problematic, however, because it seems that before the being we call Satan became Satan, he was first an angel called Lucifer, and Lucifer and the other angels apparently did have free will in Heaven, because he opposed God and enticed 1/3rd of the angels to follow him. In Peck's defense, it can be argued that once Lucifer was cast down to earth to become Satan, he (which became it) lost his/its free will at that point. There was no way Satan could change its mind, repent, or find redemption; its fate was sealed. Since it knew that, there would never be any point in trying, which is why it never has and never will, but instead will continue opposing God and his special creation (humans) until the bitter end.

This is all quite an evolution of thought for Peck, who began his writing career back in the 1970s with the best-selling book *The Road Less Traveled: A New Psychology of Love, Traditional Values, and Spiritual Growth* (Touchstone Books, 25th Anniversary Ed., 2002), in which he speculated whether belief in God was a mental illness. Even then, however, he concluded that, no, but certain beliefs *about* God are certainly destructive and/or can lead to mental illness.

Elaine Pagels, in *The Origin of Satan* (Random House, 1995), builds a historical case for the invention of the Satan character in Christian orthodoxy. She does not believe Satan is a being. She does not even seem to believe that Satan represents a force of evil. Instead, she shows how ancient Jews created the myth of "a" satan—an angel who served as an adversary to block certain human actions. She also explains that the Greek term from which we get our English word "devil" is *diabolos*, which translates literally as "one who throws something across [another] one's path" (39). According to Pagels, there was no Satan character in Jewish mythology until the 500s B. C., during and after the Babylonian captivity. It was in that troubled time that the books of Job and Numbers were written in their current form. Over time, the myth was refined and revised as the Jews interacted with Persians, Greeks, and Romans. Much of the evolution of the myth came from apocryphal writings such as the book of Enoch rather than from the orthodox books of the Old Testament. It was Christians, however, who gave final shape to the evil character we call Satan.

HEAVEN

Jeffrey Burton Russell's *A History of Heaven: The Singing Silence* (Princeton University Press, 1997) contains a wealth of, what was for me, new information on a variety of disparate topics. For example, English is the only major western language, says Russell, that distinguishes between "heaven" and "sky" (xiv). Another: our English word "ineffable" best describes the indescribable, and concepts such as Heaven, eternity, and even God, are all ineffable. The term originated in the theology of Augustine in the 4th-5th centuries as the Greek word *ineffabilis* (6). Augustine also wrote a treatise in 401 called *De Genesi ad Litteram* (*On the Literal Interpretation of Genesis*), but Russell explains that the word "literal" as used here did not mean what we take it to mean in modern English. We use "literal" to mean a factual, scientific, and/or historical meaning of a text, but Augustine used it in the original Greek sense—as an expression of what the author of a text was trying to say. So in discussing Genesis, Augustine did not read it as a factual, scientific, historical account but rather as a story with a message, and his job was to recount and expound upon that message (7). Russell coins a great one-liner on this subject when he says that "God is a poet at least as much as a scientist or a historian" (9). That means, essentially, that God revealed his message to humanity about origins through metaphors that we can understand rather than through "facts" that our limited physical brains cannot comprehend.

Russell discusses the issue that I address of where God was before He created anything. If Heaven is a "place" as opposed to a "state" of being, then God must have created it at some point, as seems to be implied in the statement, "In the beginning, God created the heaven and the earth." But how can that be? Was God simply "nowhere" prior to that? Clearly, God exists not in a "place" called heaven but in a "state" or perhaps dimension called heaven or the heavens, which we should think of as synonymous with eternity and the spirit realm. Russell goes on to explain that time as we know it is meaningless to God, in the sense that He is not bound by it or limited by it; He experiences past, present, and future simultaneously. Russell opines that this ineffable God-place-state, which he calls "heaventime" will be experienced by humans in the afterlife not as God himself knows it but only as we are capable of knowing it—in some type of sequence of events. In other words, we will still perceive "before" and "after," even though God himself has no need of these concepts (10-13).

Russell also offers yet another explanation for the Hebrew term for God's name as revealed to Moses EHYEH ASHER EHYEH. Whereas it has commonly been translated into English as "I am that I am," and other scholars cited in this bibliography have translated it "I am what I am," "I am who I am," and/or "I will be who/what I will be," Russell renders it "I AM WHO AM" (9). Having already put a good bit of thought into this topic before reading Russell's translation, I have come to conclude that the real meaning behind God's name for himself as given to Moses is "I exist" or "I am real." Moses had asked God in the burning bush (Exodus chapter 3), when pharaoh wants to know who sent me, how should I answer? God's response was designed to convey the message to Moses, pharaoh, and the whole polytheistic world of antiquity that he, and He alone, is the "real" God, even though He is the only one of the so-called gods who was invisible or had no image or statue.

Another important point from Russell's book is that the reason it seems man has created God and that different generations of Jewish writers developed the scriptures by revising and refining them to fit the times is that the Holy Spirit revealed the truth of God to humans little-by-little over centuries as we were able to receive, comprehend, and appreciate it (17). I agree. If God had just dumped the whole truth of everything there is to know upon mankind all at once, our ancestors would not have been able to comprehend it, and we today would not be able to appreciate it. The only way divine revelation can work is for it to be given in small doses over time. Take the book of Revelation as an example of why that is so. That book, by far, offers more to chew on intellectually than any other book of the Bible. It is far more mystifying than all the rest of the Bible combined, especially for those of us who read it as primarily an account of future events rather than metaphorically as a description of life within the Roman Empire in the first century when John penned it. It almost leads me to say that God gave us too much information in that one *tour de force* of the Christian scriptures. I won't say that, though, because obviously God knows what He is doing. But the point remains that, as of today, the meaning of Revelation is still mostly beyond our grasp, try as we might to understand it.

A final topic here from Russell's book is on the meanings of the Greek words *pyche*, *pneuma*, *soma*, and *sarx*. They all have related meanings with subtle differences, and often these subtleties are the thing that causes the problems of translating accurately into English what the authors were trying to say. In these four words are contained multiple concepts, such as the life force, life principle, or spark of life; breath or wind; spirit, soul, or ghost; and self or being (24). Considering how these four words have been used in various places in scripture to mean any of these things or any number of different combinations of these things, there is no wonder Judeo-Christian theology needs *The Explanation*!

CHRISTIAN INTERPRETATIONS OF GENESIS

The Unfolding Drama of Redemption: The Bible as a Whole, by W. Graham Scroggie (Zondervan Publishing House, 1970) is among the most magisterial works of its kind ever written. It covers the whole Bible from Genesis to Revelation and shows how all the books fit together and tell a unified story—the story of humanity's reconciliation to God through Christ. It offers hundreds of interesting insights on how to make sense of the Bible, and it does so more effectively than most other books of this kind because Scroggie uses layperson language rather than stilted academic jargon.

As far as his exegesis of the creation story goes, Scroggie takes it overtly rather than metaphorically. In places he admits that he doesn't know whether the term used or the concept described should be taken that way or not (such as the "tree of life"—was it an actual tree, or did that terminology represent something else?), but in such cases he simply allows for it to be a mystery and moves on. He rejects the whole notion of different authors compiling, editing, and revising Genesis, *ala* the J, E, P, and D theory. He says instead that in chapter 1 we get a summary declaration of the facts of creation, and in chapter 2 we get a description of the process of creation. Likewise, we see the creation of man in chapter 1 as the culmination of a chain of events, but we see the same in chapter 2 as the beginning of human history. He sees no contradiction in the two accounts of the creation of man but rather that they complement one another. That is how I read these chapters, too, in formulating *The Explanation*.

On the subject of the "image" of God, Scroggie says that, "Because Adam was a man and not a machine.... Man was given a will.... Without the exercise of the will Adam could not become either holy or sinful. Worth or worthlessness is determined by testing" (56). This makes sense, of course, in one of only two ways—one, if we think of God as being less than omniscient and needing to conduct a test to find out the result; or two, if we think of the test as being given not so God can find out whether his humans would be worthy or worthless but so that humans could find it out themselves. I believe God knew in advance the outcome, so there was no need for the test for his benefit. The test was for the benefit of his created beings—humans, angels, and devils. The only way these limited creatures would ever truly know and accept their less-than-God-ness would be to let them discover it for themselves.

On the subject of the serpent's temptation of Eve, Scroggie points out something that I found to be an astute observation—that the serpent acknowledges God (Elohim) in conversing with Eve but never mentions the Lord (Yahweh). To me, that means he acknowledged the spirit-Father that we commonly call "God" but not the bodily manifestation of God that we think of as Christ the Son. This leads me to ponder a point of theology and build part of *The Explanation* around it; namely that the serpent, which represented Satan, was jealous of "The Lord" (Jesus), seeing him as more of a rival perhaps than as fully God. We tend to take it for granted today that Satan always knew that Christ was God in the flesh, just as orthodox Christian theology teaches. That may not be the case at all, however. He may well have seen the bodily Yahweh (the Lord Jesus) as being no greater than himself. I can picture him sizing up Yahweh/Jesus and thinking to himself, "I bet I can beat this guy." Before you disagree, consider that once Jesus entered his earthly body as the son of Mary, Satan tempted him for 40 days in the wilderness. Surely, we assume, God the Father *allowed* this temptation of his Son, just as He allowed the temptation of Job. He did this, as I see it, not to find out whether Jesus would crack under pressure but to prove to Satan himself that he was wrong, that he couldn't beat this guy after all. In a similar vein to Scroggie's book, see the much more concise Edmund P. Clowney, *The Unfolding Mystery: Discovering Christ in the Old Testament* (P&R, 1988), and Max Lucado and Randy Frazee, *The Story: The Bible as One Continuing Story of God and His People* (Zondervan, 2011). The latter, interestingly, omits discussion of Cain's banishment and his taking a wife to the land of Nod.

GNOSTICISM

Stuart Holroyd's *The Elements of Gnosticism* (Element Books, Ltd., 1994) gives an excellent brief overview of what Gnosticism was/is. From it, I began to put some puzzle pieces together that had not quite fit together in my mind prior to that. It explains that Gnosticism was/is not just one thing, not a unified belief system, but rather has had many variations over time from one group of Gnostics to another. What Holroyd does which I found helpful to clarifying my thoughts on this subject was identify the most common elements of this diverse belief system. Once identified, these points are easily contrastable with orthodox Christian doctrines, and I suppose to other religions as well. For example, a typical Gnostic belief is that there is not one creator-God but two, one good and one bad, both termed Demiurges. The good has a team of under-Gods called Aeons, and the bad has a team called Archons. Neither side is supreme; they are locked in a struggle eternally. Another is that above these Demiurges is a single transcendent God who never "created" anything; He never had to. He merely thought about a thing, and it came into being. To say He "thought" doesn't do justice to the idea I'm trying to convey, however, because thinking is a type of action, and this God doesn't need to act. Perhaps a better way to put it is that his will results in whatever He wants, and it is automatic and instantaneous. From his will, certain things emanated, such as thought, wisdom, truth. And I suppose that their equal opposites—emotional reaction, foolishness, and falsehood—came into being as a consequence (although this book doesn't say that exactly; I'm extrapolating here). He is not concerned with the Demiurges' acts of physical creation. He is above all of that—always has been and always will be.

This all sounds similar to Greek and Roman mythology to me, with important variations, of course. The template is close in that it shows an eternal struggle between good and evil playing out in the heavens just as it does here on earth. The one thing about this Gnostic belief system that seems similar to the Judeo-Christian version of creation is the concept of a transcendent, non-physical God at the top. Beyond that, I believe the Gnostics were and still are reaching for the explanation but are starting from a different base line than I am. I start with Genesis as being the inspired Word of the transcendent God and therefore the main piece of the puzzle that all other pieces must fit around. They start with who-knows-what—ancient Egyptian religious ideas and/or other ancient mystery religion ideas and build upon them or vary them from culture to culture, generation to generation; Zoroastrianism is one example. I think that the Satan character of the Bible is responsible for deceiving early humans into starting these mystery religions, but I think the "gods" or "Demiurges" of these ancient religions were actual physical beings (what we Christians would call angels, fallen angels, and/or demons; and what the secular ancient alien theorists would call extraterrestrials).

Richard Smoley's *Forbidden Faith: The Gnostic Legacy from the Gospels to 'The Da Vinci Code'* (HarperSanFrancisco, 2006) also gave me some important insights, but they are of a different variety than those of the Holroyd book. This one points out certain problems with orthodox Christianity and shows how adopting a Gnostic version of Christianity might solve them, or at least offer an alternative. Take the doctrine of salvation, for instance. Smoley says the orthodox teaching about it contains a great amount of truth, because it:

> provides a powerful answer to the universal problem of human guilt. But to say it has truth is not to say it is the whole truth. . . . Salvation, if it is genuine, clears the boards, enabling one to go on to lead a meaningful and purposeful life. But salvation itself does not confer meaning or purpose; it is merely one step in that direction. . . . To believe that sin and redemption constitute the whole of the spiritual path puts too much focus on sin. This is the trap into which conventional Christianity has fallen. The believer continues to lament and beat his breast because, as it were, he has nothing else to do (206-207).

I agree that, unfortunately, this seems to be the case with a lot of Christians and with certain denominations which emphasize repentance and redemption so much to prepare us for eternity in Heaven that they never get to the abundant life part of the Gospel which takes place in the here-and-now. I also realize that this is a stereotype and an oversimplification of Christianity in general which does not hold true for a large percentage of the whole. Smoley goes on to explain in so many words that while Christianity showcases unconditional love as the ultimate thing to aspire to as human beings, Gnosticism offers detachment from self, the physical world, and normal consciousness as the ultimate thing. This is, by his own admission, an Eastern religious ideology rather than a western one. Gnosticism, therefore, bridges the gap between the East and the West, religiously speaking.

That concept was useful to me as I continued formulating *The Explanation*, because I argue that if we could go back in time to the populating of the earth after the Tower of Babel incident in Genesis, we would see the root of all religions—Eastern, Western, and anything else (if there is anything else). The variety found in Gnosticism ever since has resulted merely from each group of people putting their own twist on that original Satanic religion as they spread to different parts of the earth. That is why there are so many similarities between ancient cultures on different continents that seemingly had no knowledge of or connection with one another—they all originated from a common source.

Smoley tackles another aspect of Christianity herein, too: that of Evangelical Christians claiming to have "a personal relationship with Jesus Christ." He asks what exactly it is to have a personal relationship with an "Imaginary Friend." He follows that by asking what we mean when we say we have been "born again." Basically, he lumps both the one-time born again experience together with the ongoing personal relationship with Jesus and describes them thusly:

> The believer creates a mental picture of Jesus from what he knows from reading the Bible or going to church, the mind then makes this picture come to life (which it can easily do), and then he has a relationship with this figment.... If by some chance the individual has a genuine experience of the divine, he only accepts it if it fits in with his preconceptions. Otherwise he hates and fears it as a snare of the devil.... This is not to deny that a person can have a genuine encounter with the presence of Christ, which in itself is a form of gnosis. But many people appear to have had "born again" experiences largely because they were expected to. They soon discovered that these experiences were empty. Or, just as often, they found that the experience itself was powerful, but their churches had hedged around it with so many unrealistic moral, doctrinal, and even political demands that they become disillusioned and embittered. Disbelieving in the church, they came to disbelieve in their own experience (208-209).

I can only say "amen" to this block quote. I say much the same thing in *The Explanation*, at least concerning the image that we humans construct in our minds about what God looks like and/or what his nature is. The part about disillusionment with organized religion causing people to fall away from their own inner faith may not be stated in *The Explanation*, but it is certainly an important and valid observation worth mentioning here.

SCIENCE, GENERALLY

Because I am not a scientist and make no pretense of understanding the esoteric, specialized jargon used by scientists in all of the different fields, much less sub-fields, I will lump together a bunch of books that I looked at and gleaned information from. It, like many of the things I have written in *The Explanation* on scientific subjects, will positively insult the intelligence of professional scientists, but again, I must remind readers that impressing such scholars (or avoiding insulting them) is not my goal in this study. So bear with me.

Frederick S. Szalay and Eric Delson's *Evolutionary History of the Primates* (Academic Press, Inc., 1979) is a paleobiological study. The authors take pride in using their specialized knowledge of primate evolution to counter more popular works written by anthropologists. They point out that whereas anthropologists are concerned with how primates (supposedly) evolved into humans, paleobiologists are not. They are concerned rather with non-human primate evolution into other non-human primate types. The main point I got from this book was the fact that scientists in one sub-specialty can be quite territorial and judgmental, just as Mary Jane West-Eberhard (who I mentioned at the beginning of this bibliography) pointed out. They do not generally like to be challenged in matters which they spend their lives and careers researching, teaching, and writing about; hence, the antipathy scientists in general feel toward theological attempts to explain natural phenomena. Concerning anthropological works that fit the mold that Szalay and Delson describe, J. Bronowski's *The Ascent of Man* (Little, Brown and Company, 1973) is one that I looked at. It discusses the supposed origin of humans in the Omo valley of Ethiopia, yet it also says that Middle Eastern Neanderthals probably ascended directly into homo sapiens.

Many science books were instructive to me on just one point that I raise somewhere in *The Explanation*. M. J. D. White, in *Animal Cryptology and Evolution*, 3rd ed., (Cambridge University Press, 1973), pointed out something that I knew and took for granted in formulating my thoughts on the possibility that Satan and other fallen angels had creative power—viruses don't have DNA; they have only RNA. Without DNA, there is no life in what would otherwise appear to be a living organism, such as a virus, which can move and replicate itself only if it has a living host to attach to. Stephen R. Anderson, meanwhile, in *Doctor Dolittle's Delusion: Animals and the Uniqueness of Human Language* (Yale University Press, 2004), points out in the very title of his book that humans, among all species on earth, have the ability to communicate with an actual verbal language, which is important in supporting some of my thoughts on the serpent/snake's ability to talk in the Garden of Eden. Likewise, Barbara Herrnstein Smith's *Scandalous Knowledge: Science, Truth and the Human* (Duke University Press, 2005) contains a chapter called "Animal Relatives, Difficult Relations," which discusses the issue that I raise in *The Explanation* of why humans generally care for the welfare of certain animals but not others. Why do most people love dogs, for example, but hate snakes? Why do people tend to love mammals but feel no great affection for, say, insects?

Meanwhile, Richard W. Burkhardt, Jr.'s *Patterns of Behavior: Konrad Lorenz, Niko Tinbergen, and the Founding of Ethology* (University of Chicago Press, 2005), was instructive to me for its history of the subfield of Biology called Ethology, which is the study of animal behavior.

Although there had been scientists studying animal behavior as a specialty since the late 1800s, Ethology as a separate, professional field of study was only recognized in 1973. The field is devoted to understanding and cataloguing the instinctive behaviors of all the different species. Why birds fly south for the winter, for example, is the type of issue Ethologists would be concerned with. Why dogs can be "man's best friend" yet were not found worthy of being Adam's helper in the Garden—not so much.

Some science books stood out among the crowd to me. George C. Williams's *Adaptation and Natural Selection: A Critique of Some Current Evolutionary Thought* (Princeton University Press, 1966) seemed remarkable to me for its succinct way of explaining a major point concerning how secular scientists typically frame their arguments in favor of godless evolution. Ironically, despite being a major point, and despite the fact that Williams brought it to the attention of the scientific community in the mid-1960s, it seems to be often deliberately ignored still today. It is this: natural selection can be observed only as an "effect," meaning we only see the results of it after it has happened, yet upon seeing those results or effects, scientists tend to speak and write of the "function" of the mutation as giving an "advantage" to the member(s) of the species which have it (252-255). Terminology here is important: function and advantage imply design or purpose, even if the scientists who use those terms don't intend the words to be taken that way. True naturalists (atheists or those who simply leave God out of the equation because no supernatural agents are allowed in natural science) should never even hint at their being a causal agent at work on the front end of the natural selection process, says Williams, but they do it all the time without realizing it. A related logical fallacy that routinely gets employed by evolutionary scientists is the implied assumption that each species has some kind of "collective interest" in ensuring the survival of the species (253). Except for homo sapiens, that cannot be, says Williams, in a purely naturalistic universe where each individual is concerned with its own survival and reproduction, not that of the rest of the species. For a whole species to have collective will to live and reproduce somehow smacks of design. On this same subject, see also J. Merritt Emlen, *Ecology: An Evolutionary Approach* (Addison-Wesley Publishing Company, 1973).

Williams goes on to question, speculate upon, and philosophize about why his fellow naturalists fall victim to this logical fallacy. He decides that it is probably because of these scientists growing up in, and their field of study developing within, Judeo-Christian culture. He wonders how much more "mature" the field would be (as of 1966) if the scientists and their field could have come of age in a culture that wasn't permeated with religious concepts (255).

For me, reading this re-circulated a question that has been rolling around in my head for several years: why is it that most of the advances in science which have produced the modern, technological world that we live in come from the so-called "Christian" nations? This is a fact, regardless of how much some may want to deny it or spin it. One need look only at the *World Almanac* (any one of many recent editions) to see a list of some of the most important discoveries, inventions, and innovations in our modern world to see where the men (almost all were/are male) lived when they made their great contribution to science. The United States and Great Britain dominate, but the rest of the nations of

Europe are about evenly divided, while scarcely any great development came from elsewhere.

Speaking of the philosophy of science, I looked at several books in this category. One was Lisa Sowle Cahill, ed., *Genetics, Theology, and Ethics: An Interdisciplinary Conversation* (Herder & Herder Books, 2005). It debates whether humans can or should be "co-creators" with God (2). If humans can alleviate suffering through advanced technology, is it always ethical to do so? Or does alleviating suffering require trade-offs? Stem cell research, genetically designing children before conception, cloning, and abortion are all examples of questions that blur the lines between science and theology and bring ethical considerations front and center. Before godless scientists spring into action to see how far their research can take them on the road to becoming co-creators with God, they would do well to remember a point brought out powerfully by David H. Smith and Cynthia B. Cohen, eds., in *A Christian Response to the New Genetics* (Rowman & Littlefield Publishers, Inc., 2003): "it is the height of prejudice to assume that one can learn nothing from traditions that have sustained persons and groups for millennia" (ix), meaning, basically, from the Bible and the Christian faith.

PREHISTORIC ARCHAEOLOGY AND ANTHROPOLOGY

There often seems to be a murky distinction between what historians, archaeologists, and anthropologists do. Their fields overlap in some cases. As a historian myself, I can generally tell when I'm reading a fellow professionally trained historian's work, but archaeologists and anthropologists (not to mention philosophically-inclined scholars in other sciences, social sciences, and humanities fields) are often hard for me to distinguish between. It is helpful when they identify themselves and their professional specialty clearly at the beginning of their books. It is most definitely not helpful to find archaeology books shelved amidst history books, or anthropology books shelved amid zoology books. I encountered some of that in my research. It seems to me that, technically speaking, archaeologists ought only to do the digging and let anthropologists and/or historians do the interpreting of their findings. Most archaeologists, I feel sure, will not appreciate my opinion on that, but it would nonetheless help clear up some of the overlap that currently exists if that were the case.

From John Dimick's *Episodes in Archaeology: Bit Parts in Big Dramas* (Barre, 1968), I learned the original *Webster's Dictionary* definition of archaeology from 1928: "The study of material remains of past human life and human activities" (1). V. Gordon Childe's *Progress and Archaeology* (Watts & Co., 1944) is a short little survey of the field as of the 1940s. The author implies that archaeology is a type or part of history. I say it is actually a field that is largely devoted to prehistory and the non-historical past, however, meaning the past for which there are no written records. This author influenced a whole generation of students, so his choices and uses of words mattered at the time. Interestingly in his book *Man Makes Himself* (A Mentor Book, 1951), he approaches the subject strictly as prehistory rather than history, and in his 1956 book *A Short Introduction to Archaeology* (Collier Books), he calls archaeology a "source" of history (9). In his aforementioned 1944 book, he did a great service, however, in accurately defining three words that are all-too-often vaguely understood when not altogether misunderstood:

"savages" means exclusively hunter-gatherers; "barbarians" means those who farm and raise livestock in addition to hunting and gathering; and "civilized" people means those who live in cities and have a division of occupations and labor within their society. This all applies to *The Explanation* with regard to my discussion of what happened when Adam and Eve got expelled from the garden and when Cain got expelled from Eden. They became "savages" initially, but by the time Cain killed Abel, the family had clearly progressed to the stage of "barbarians." Not until there were enough people on earth to justify building cities did these early humans become "civilized."

Childe also points out that there was an ongoing feud between the "evolutionists" and "diffusionists" within the archaeological/anthropological community. The former believed that each society which arose anywhere in the world would eventually discover or invent the same new things given the availability of similar resources. Diffusionists, however, believed that each discovery or invention was made only once and then spread to all other societies in due course. (Clearly, since the time this book was published, the diffusionists won out). Either way, he says the earliest evidence of the building of temples and the performance of sacrifices or offerings to gods dates back to only about 4,000 B. C., despite the fact that archaeological remains of other types date back to about 340,000 B. C. The 4,000 B. C. date is easily verifiable, and it is corroborated by the Bible, but where the 340,000 B. C. time span comes from, I don't know. He doesn't say. I am skeptical of it.

Noel Cowan revises that 340,000 year time span to about 100,000 years in his *Global History: A Short Overview* (Polity, 2001). He then gives early humans a 90,000 year window to overspread the world from their supposed East Africa point of origin. He explains that genetic data which had only recently become available proved the diffusion theory and disproved the simultaneous evolution theory. He proffers the standard Bering land bridge theory to show why the Americas seem to have been the last continents on earth to be inhabited by humans.

Murray T. Pringle's *Wonders Under the Earth* (Lantern Press Inc., 1966) does a service by defining history fairly accurately (not bad for an archaeologist): the study of "the written record of man's achievements" (11). Written records are the key to history, which is why all the bones, pottery, remains of cities, etc., that archaeologists dig up are not history. Pringle notes that history dates back only about two and a half millennia, and writing itself dates back only about twice that long, or so the archaeological evidence suggests. Of special importance to me in this book is the author's brief discussion of the Toltec civilization in Mexico, including its great city of Teotihuacan, and its god Quetzalcoatl, who Toltec writers called a bearded white man from the east. This "god" taught the people advanced mathematics, astronomy, architecture, agriculture, government, and recreational sports. To me this leads to only one conclusion: this "god" was no mythical being concocted by the people but was rather an actual, physical humanoid creature with advanced knowledge and technology that once lived among them. This is the ancient alien theory in microcosm.

Sir Leonard Woolley discusses a dig at a site called "Nimrud," which he identifies as the biblical city of Calah, in *History Unearthed* (Ernest Benn Limited, 1968). In Iraq today, it was an Assyrian city that has no relation to the man named Nimrod in Genesis Chapter 10. He also discusses an archaeological dispute over how to interpret the evidence unearthed in a dig in Ur, a Sumerian city mentioned in Genesis. He believed he had found a tomb of kings and queens, but other scholars disagreed because it conflicted with their conventional wisdom. He discusses a dig at Jericho, too, which he believes had already been conquered by Egyptians some 200 years before the Israelites arrived to knock down its walls.

Alice Beck Kehoe discusses this same topic of Jericho, along with several other biblical topics, in her *Controversies in Archaeology* (Left Coast Press, 2008). She points out that once a belief about the past takes root in popular culture, it is nearly impossible to dislodge. She cites Stonehenge as a prime example. For centuries it was assumed to have been a Druid site, but we now know it predated the Druids by a couple-thousand years. Even so, Druid re-enactors converge on it every year to recreate their make-believe version of the past. About dislodging belief in the Bible, she takes it for granted that no archaeologist worth his salt today actually believes it is anything more than a collection of myths, even though many of the sites and some of the people mentioned in it have been found to be true. Those secular archaeologists who specialize in biblical digs are called "Syro-Palestinians." The miracles mentioned throughout the Bible are completely unbelievable to her. One important piece of information I learned from this book is that in 1844, a Scottish journalist named Robert Chambers published *Vestiges of the Natural History of Creation*, which put forth a pre-Darwin evolution thesis. It started with a sort of Big Bang cosmology and traced the origins of earth and its life forms as ever more complex things building upon simple things.

Several books in this genre yielded isolated shards of information that I found useful, although unconnected to one another. Paul J. Alexander, ed., said in *The Ancient World: To 300 A. D.* (The Macmillan Company, 1963) that Hammurabi was ordered to write his code by the sun god Shamash. I had previously read in other sources that the god was Marduk. He also showed that the Akkadians had their own creation myth poem dating back at least to the first millennium B. C., although some scholars believe it is about a thousand years older than that, placing it perhaps around 1,700-1,800 B. C. The Epic of Gilgamesh, he implies, is really just a continuation of that same story, much like the story of Noah is a continuation of the story of Adam and Eve in Genesis. Regardless, he says that the original version of the Epic of Gilgamesh is written in Akkadian, not Sumerian. I had always gathered from other sources that it was Sumerian, and this raised questions in my mind: what exactly is the difference between the Akkadian and Sumerian labels? Does the label denote two separate languages, or two separate nations or ethnic groups, or both? Or do the labels identify the same basic group of people who just occupied different areas of Mesopotamia at the same time? Or was one group the forerunner of the other but otherwise basically the same people?

Some convenient internet research cleared up the confusion for me, although this is actually quite a complex topic. It turns out that "Sumerian" refers mainly to a language, not so much a nation or ethnic identity, although it obviously grew out of the city-state of Sumer in southern Mesopotamia. Akkadian refers to language as well, but in history the term came to represent a national identity—an empire—based on a specific group of people who controlled northern

Mesopotamia at a particular time. Although both Sumer and Akkad existed at roughly the same time and enjoyed some overlap, Sumer was older, and the Akkadians borrowed from the Sumerians' language and culture, including material that went into their so-called "standard" version of the Epic of Gilgamesh. So, to say that The Epic of Gilgamesh is Sumerian in origin would be true, but to say that the version of it in its entirety that has become the best known and most often cited one is Akkadian would also be true. To me, therefore, it is actually much ado about nothing. Only scholars who specialize in this little subfield of history need be concerned about such a distinction.

From R. R. Marett's *Head, Heart, & Hands in Human Evolution* (Henry Holt and Company, 1935), I learned that "anthropology" means literally "talking about people" (11). I also learned (if indeed this is true) that Herbert Spencer coined the term "evolution." It seems to me that the term should predate Spencer, but maybe not. Marett states as a matter of fact that the earliest humans thought in concrete ideas rather than abstractions. If so, this is one piece of evidence in favor of the earliest humans seeing and interacting with physical "gods" rather than making up imaginary gods as mainstream history claims. From Grahame Clark's *Aspects of Prehistory* (University of California Press, 1970), I learned that archaeology as a professional discipline really dates back only to about the 1500s, but not until the Enlightenment did it begin to be popular enough to make a splash in academia. See also David Rodnick, *An Introduction to Man and His Development* (Appleton-Century-Crofts, 1966) for a standard secular evolutionist's timeline of events from the Big Bang to the emergence of modern homo sapiens, and Alexander A. Goldenweiser, *Early Civilization: An*

Introduction to Anthropology (Alfred A. Knopf, 1922) for an example of just how much the notion of white supremacy (Social Darwinist thought) was an issue for academia to wrestle with in the 1920s.

THE ANCIENT ALIEN THEORY

Last but not least, let's tackle the "ancient alien" or "ancient astronaut" theory. Erich von Daniken's *Chariots of the Gods?* burst on the scene in the late 1960s and took the world by storm. Reprinted and updated multiple times, it has spawned a whole industry within the conspiracy theory subculture. (One recent paperback edition is published by Berkley Books, 1999.) It is important to note that Von Daniken did not invent the "ancient astronaut" theory, but in fact plagiarized some of the work of earlier authors. Even so, he popularized this alternative theory like no one else and is thus seen as the foremost figure in the field. The ancient astronaut theory is a heresy in the opinion of the orthodox purveyors of this field of study that might best be described as astronomical history or historical anthropology. E. C. Krupp, in *Echoes of the Ancient Skies: The Astronomy of Lost Civilizations* (Harper & Row, 1983), for instance, dismisses the heresy outright as being unworthy of any serious discussion.

In surveying the many books that fall into this category, I could cite a long list of titles from certain publishers, such as Bear & Company, a subsidiary of Inner Traditions. Bear & Company's catalogue includes thirteen books by Zecharia Sitchin, who, other than Von Daniken, has probably been the most influential figure in ancient alien subculture, ever since publishing *The 12th Planet* (Stein and Day) in 1976. Robert Baual and Thomas Brophy's *Black Genesis: The Prehistoric*

Origins of Ancient Egypt (2011) is another important work by this publisher, as is Christopher Dunn's *The Giza Power Plant: Technologies of Ancient Egypt* (1995). Some books that have made a big impression in this category but are published by different companies include *Technology of the Gods: The Incredible Sciences of the Ancients* (Adventures Unlimited Press, 1999) by David Childress; Philip Coppens' *The Ancient Alien Question: A New Inquiry into the Existence, Evidence, and Influence of Ancient Visitors* (New Page Books, 2011); and Graham Hancock's *Fingerprints of the Gods* (Crown, 1995).

Chris Putnam and Thomas Horn's *Exo-Vaticana: Petrus Romanus, Project L.U.C.I.F.E.R., and the Vatican's Astonishing Plan for the Arrival of an Alien Savior* (Defender, 2013) is a book with a "wow" factor off the charts. Its thesis is contained in the subtitle. The notion that the Roman Catholic Church is looking for and expecting an extraterrestrial being to come to earth and become the head of the human race seems outlandish at first glance when we think in conventional theological terms, but it makes perfect sense if we read the Bible as I do, unconventionally yet common-sensibly. The Bible is full of other-worldly beings that either come to earth from the heavens or interact with earthlings from the heavens. When Jesus departed from earth after his resurrection, He was taken up into a cloud in the sky and disappeared. Mark 16: 19 says He was *"received* up into heaven," and Luke 24:51 says He was *"carried* up into heaven." The book of Acts is the continuation of Luke's version of the Gospel, and in chapter 1, verse 9, it tells us that Jesus was "taken up; and a cloud received him out of their sight." If this is not a description of what we would recognize as an alien encounter, a UFO sighting, or an otherwise unexplainable extraterrestrial event, then what is? If this description were given in any other context besides the Bible—especially in a modern-day setting—there would be no other conclusion to reach than that we are dealing with something from outer space. For some reason, when we cloak this kind of thing with a religious veneer, we seem to think it is something different than what it clearly would be in any similar, secular case.

Anyway, Putnam and Horn are not the first ones to equate biblical accounts with extraterrestrial phenomena, but they are on the cutting edge of plumbing the depths of the subject. Those who are not willing to break free of conventional thinking on this subject will dismiss this book as a gigantic conspiracy theory. Those with an open mind will find much to ponder here, however. From demonic genetic experimentation on human abductees to the Vatican's complicity in keeping its contact with aliens secret until the appointed time for the revelation, this stuff is frankly better than fiction. It stretches the limits of one's imagination, and for that, whether it is all true or not, it demands our consideration.

MISCELLANEOUS, RANDOM, AND WORTHY OF MENTION

- Ronald Hendel's *The Book of Genesis: A Biography* (Princeton University Press, 2013) strikes me as standard scholarly academic fare. It was useful to me in that it documents the history of the critical analysis of Genesis by scientifically-minded scholars over the centuries. Hendel credits Spinoza with being the first person to write about Genesis from an objective, critical, scientific point of view rather than a believer's theological point of view. Or at least Spinoza was the first to get away with writing about it that way. Certain others who tried were burned at the stake or otherwise persecuted by the Catholic Church. What Spinoza did, says Hendel, was make a distinction between the "sense" of the text and the literal "truth" of the text. Making this distinction became known as practicing the "historical-critical method," and it was a logical extension of the larger scientific method which began sweeping through Europe during the Renaissance and Enlightenment eras. Spinoza's work opened the door for later scholars to develop the JEPD model of Genesis's authorship. Hendel also edited *Reading Genesis: Ten Methods* (Cambridge University Press, 2010), a collection of articles written by scholars for scholars. It is largely a work of historiography and literary criticism.

- In Robert Alter's *Genesis: Translation and Commentary* (W. W. Norton & Company, 1996), I saw something that struck me as unusual. Alter finds value in the original King James translation and the Tyndale translation because they preserve many Hebrew idioms even when they don't necessarily make for the most meaningful words, phrases, or sentences in English. Implied, therefore, is that most other translations put more emphasis on making meaningful English verses than on being faithful to the Hebrew. I have rarely seen the KJV defended in my research thus far, so this is refreshing. Most of the rest of this book is standard fare, but I did learn that the cherubim which guarded the way of the tree of life should not be visualized as beautiful, humanoid angels as they generally are portrayed by artists, but rather as hybrid flying creatures. Alter notes that such sky creatures were gods in Canaanite mythology. I draw from this more ammunition for the ancient alien theory—what early humans mistook for gods were highly advanced other-worldly beings.

- E. Basil Jackson's *The Faith Dynamic: A Treatise on Creationism and the Evolutionary Theory* (Post-Gutenberg Books, 2013) is written by one of the most highly educated people to appear in this bibliography. He holds an MD, a JD, and a Ph.D. He also has one of the most international backgrounds of anyone on this list, having been born in Northern Ireland, receiving his Ph. D. in South Africa, and practicing law and medicine in the United States. He admits that he has an "affinity" for the Reformed Baptist tradition, and consequently that his perception of the issues herein may be colored by his Christian faith, yet his strong secular education in psychiatric medicine and forensic law mitigate, if not altogether negate, any bias that might be present in his writing. In short, this book is very objective. The issue in question in this study is whether belief in evolution is as much a faith-based system as is any religious belief such as creationism. Or, to put it another way, it is whether belief in evolution is a secular religion in itself. The author does not take sides but rather presents evidence for and against the hypothesis. In the end, he determines that the theory of evolution is just as dependent on faith, albeit a different kind of faith perhaps, as is creationism, but that doesn't qualify it to be called a type of "religion" *per se*.

- E. A. Speiser, in *The Anchor Bible Genesis, Volume I: Introduction, Translation, and Notes* (Doubleday & Company, Inc., 1962), brought to my attention for (I think) the second time that "Elohim" is a plural noun in the ancient Hebrew. This, of course, lends credence to the Christian concept of the Trinity. However, Speiser notes that "Elohim" can also mean "gods," as if implying the Hebrews or Jews had an equivalent to all the pagan gods of the ancient world, not necessarily an omnipotent God above all other so-called gods.

- Edwin M. Good, in *Genesis I-II: Tales of the Earliest World* (Stanford University Press, 2011), tries to make meaningful translations of ancient Hebrew texts of Genesis into our everyday English. Instead of using "God" in his commentary on Genesis Chapter 1, he uses "Elohim" exclusively. Starting in verse 4 of Chapter 2, he uses "Yahweh Elohim" exclusively. In each case, he does so because it makes for the most meaningful translation to keep these Hebrew terms as proper names rather than to debase them into the generic "God" and "Lord God." This seems helpful to me in confirming my belief that Elohim represents the spirit-God in Heaven and Yahweh Elohim represents the physical version of God that comes to earth to interact with his creation. Good does not make that connection, of course, but then, I wouldn't expect him or anyone else to. I will be surprised if in the course of my research I stumble upon someone else who already came to that conclusion. If so, I will bring my false assumption to the reader's attention herein and give my sincerest apologies. Good does point out several of the same problems with the Genesis narrative that I point out in *The Explanation*. One example is that God finished his work on the seventh day, not the sixth, and then rested on the seventh. He punts on this point, however, as I have noticed many otherwise excellent scholars do on many points in commenting on Genesis, by urging readers not to worry about the inconsistency. His interpretations are in some cases starkly different than mine, and I dare say his are more standard and mine more unorthodox.

- R. Crumb's *The Book of Genesis Illustrated* (W. W. Norton & Company, 2009) is a most interesting book. Just because it is a picture book does not mean it is a children's book. It tells the story of Genesis by providing an illustration and caption for every event, conversation, and scene. The illustrations have adult themes, so wherever sex is mentioned in Genesis, there is a picture of it. It is not pornographic, however. It is extremely modest in depicting what the narrator of Genesis tells us happened. Although the author tells us right up front that he does not believe the Bible is the inspired word of God, his rendition of the story is faithful to the actual text of the Bible. He claims to have taken no liberties with the story, but there is a confusing passage where he discusses Abraham and

Melchizedek (Genesis Chapter 14), and goes back and forth from the King of Salem to the King of Sodom.

- Carlos Mesters' *Eden: Golden Age or Goad to Action?* (Orbis Books, 1974) is the American version of the Brazilian book first published in 1971. Mesters makes some important points, although I don't agree with all of them. One is that we cannot know what the original sin was that Adam and Eve committed, and to worry about it or inquire too deeply into it is a waste of time that misses the point. I disagree. He asks the question, "Why didn't God give Adam and Eve another chance?" That, I believe, is a question based on the false premise that God punished Adam and Eve for sinning, when actually God merely informed them of the consequences of their action. Another is that the serpent was a Canaanite religious symbol representing evil, and that's where the author of Genesis got the idea for the serpent from. I disagree. There may be a tie-in here, however, with my belief that pagans worshiped fallen angels, and that one of them took the form of a serpent or dragon.

- Peter Coveney and Roger Highfield, in *The Arrow of Time: A Voyage through Science to Solve Time's Greatest Mystery* (Fawcett Columbine, 1990), claim there is some observable laboratory evidence of order spontaneously "returning" after a period of chaos.

- Peter Shockey, in *Reflections of Heaven: A Millennial Odyssey of Miracles, Angels, and the Afterlife* (Doubleday, 1999), postulates the quantum physics theory that God is pure energy, which manifests in movement of particles throughout the universe and in the emission of light as a by-product.

- Elaine Pagels and Karen L. King's *Reading Judas: The Gospel of Judas and the Shaping of Christianity* (Viking, 2007) contains an informative discussion about what constitutes the "image" of God.

- Daniel Berrigan's *Genesis: Fair Beginnings, Then Foul* (Rowman & Littlefield Publishers, Inc., 2006) reads like poetry rather than prose. It is chopped up into hundreds of small sections, each ostensibly able to stand alone as a commentary on some verse, phrase, word, or thought within Genesis. It reminds me of Will Campbell's style of writing. Campbell was prone to make commentary on whatever the topic in question was by alluding to some obscure thing he had read or to some random thought that ran through his mind.

- Bill Moyers, along with Betty Sue Flowers, edited *Genesis: A Living Conversation* (Doubleday, 1996). This is an unusual sort of book. It features more than 30 experts holding informal conversations among themselves about how to interpret various parts of Genesis. They address ten different topics. Not all of the experts participated in every conversation.

- Harris Lenowitz and Charles Doria's *Origins: Creation Texts from the Ancient Mediterranean* (Arno Press, Inc., 1976) contains dozens of ancient manuscript fragments and excerpts translated into English and arranged as poetic verse.

- James N. Gardner's *Biocosm: The New Scientific Theory of Evolution: Intelligent Life Is the Architect of the Universe* (Inner Ocean, 2003) gives a good and recent overview of what the title implies.

- Roger Pilkington's *In the Beginning: The Story of Creation* (St. Martin's Press, 1957) is a children's book that is pro-creationism in orientation.

- Gerhard Von Rad, *Genesis: A Commentary* (Westminster Press, 1961). This is the American edition of a German book first published in 1956, translated by John H. Marks.

- Bill W. Tillery's *Physical Science*, Second Ed. (Wm. C. Brown Publisher, 1993) is the college survey textbook from which I learned many of the rudiments of science and the scientific method, while *General Zoology*, Sixth Ed. (McGraw-Hill Book Company, 1979) by Tracy I. Storer, et al., is the college survey textbook from which I learned some of the basics of what I know about zoology. Meanwhile, H. J. de Blij and Peter O. Muller's *Geography: Realms, Regions, and Concepts*, Updated and Revised Eighth Ed. (John Wiley & Sons, Inc., 1998) is a college textbook from which I learned some of what I know about geography.

SOME ADDITIONAL SOURCES CONSULTED

John Bressler, unpublished manuscript, prepared for teaching the Bible in one year at First Presbyterian Church, Statesboro, Georgia, circa 1990.

H. Lee Cheek, Jr., "Recovering Moses: The Contribution of Eric Voegelin and Contemporary Political Science," in Hebraic Political Studies, 1:4 (Summer 2006), 493-509.

Mickey L. Mattox, "Genesis 1-3 as a Problem Text in the History of Exegesis," TSF Presentation, 1992.

Dean H. Hamer, *The God Gene: How Faith Is Hardwired into Our Genes* (Doubleday, 2004).

Andrew Newberg and Eugene D'Aquili, *Why God Won't Go Away: Brain Science and the Biology of Belief* (Ballantine Books, 2001).

Bart D. Erhman, *Lost Scriptures: Books that did Not Make It into the New Testament* and *Lost Christianities: The Battles for Scripture and the Faiths We Never Knew* (both Oxford University Press, 2003).

Ryan Sharp, " If God Made the Universe, Then Who Made God?" in the *Apologetics Study Bible for Students* (Holman Bible Publishers, 2009).

Sidney Greidanus, *Preaching Christ from Genesis: Foundations for Expository Sermon* (William B. Eerdmans Publishing Company, 2007).

Michael James Williams, *Deception in Genesis: An Investigation into the Morality of a Unique Biblical Phenomenon* (Peter Lang, 2001).

Steve Jones, *The Serpent's Promise: The Bible Retold as Science* (Doubleday Canada, 2013).

Edward Caudill, *Intelligently Designed: How Creationists Built the Campaign against Evolution* (University of Illinois Press, 2013).

William P. Brown, *The Seven Pillars of Creation: The Bible, Science, and the Ecology of Wonder* (Oxford University Press, 2010).

Walter M. Fitch, *The Three Failures of Creationism: Logic, Rhetoric, and Science* (University of California Press, 2012).

Paul Davies, *Cosmic Jackpot: Why Our Universe Is Just Right for Life* (Houghton Mifflin Company, 2007).

Jon D. Levenson, *Creation and the Persistence of Evil: The Jewish Drama of Divine Omnipotence* (reprint, Princeton University Press, 1994).

Roland Mushat Frye, ed., *Is God a Creationist?: The Religious Case against Creation-Science* (Charles Scribner's Sons, 1983).

T. J. Wray and Gregory Mobley, *The Birth of Satan: Tracing the Devil's Biblical Roots* (Palgrave McMillan, 2005).

Dorothy Nelkin, *The Creation Controversy: Science or Scripture in the Schools* (W. W. Norton & Company, 1982).

Frederick H. Stitt, *Adam to Ahab: Myth and History in the Bible* (Paragon House, 2005).

Stephen C. Barton and David Wilkinson, eds., *Reading Genesis after Darwin* (Oxford University Press, 2009).

Conor Cunningham, *Darwin's Pious Idea: Why the Ultra-Darwinists and Creationists Both Get It Wrong* (William B. Eerdmans Publishing Company, 2010).

Jason Rosenhouse, *Among the Creationists: Disptaches from the Anti-Evolutionist Front Line* (Oxford University Press, 2012).

Yochanan Muffs, *The Personhood of God: Biblical Theology, Human Faith, and the Divine Image* (Jewish Lights Publishing, 2005).

James L. Crenshaw, *Defending God: Biblical Responses to the Problem of Evil* (Oxford University Press, 2005).

Alexander Heidel, *Babylonian Genesis: The Story of Creation* (University of Chicago Press, 1942).

Dan Lioy, *The Search for Ultimate Reality: Intertextuality between the Genesis and Johannine Prologues* (Peter Lang, 2005).

Rupert Matthews, et al., *Unseen World: The Science, Theories, and Phenomena behind Paranormal Events* (The Reader's Digest Corporation, Inc., 2008).

Norman Shine, *Numerology: Your Character and Future Revealed in Numbers* (Fireside, 1994).

Alma E. Guiness, et. al., eds., *Reader's Digest Mysteries of the Bible: The Enduring Questions of the Scriptures* (The Reader's Digest Association, Inc., 1988).

Ralph D. Winters, et al, eds. *Ancient World: Creation to 400 B.C., Lesson Overviews* (Institute of International Studies, 2008).

Neil deGrasse Tyson, host, "Where Did We Come From?" PBS television, *Nova*, "Evolution" Series, February 16, 2011 (transcript at http://www.pbs.org/wgbh/nova/evolution/where-did-we-come-from.html).

Hiebert, Theodore "The Tower of Babel and the Origin of the World's Cultures," *Journal of Biblical Literature* 126 (2007): 29–58.

Brent A. Strawn, "Holes in the Tower of Babel," *Oxford Biblical Studies Online*, (http://global.oup.com/obso/focus/focus_on_towerbabel/#bibliography).

Dave Lingston, "Who Was Nimrod?" (http://davelivingston.com/nimrod.htm).

Eugene E. Harris, *Ancestors in Our Genome: The New Science of Human Evolution* (Oxford University Press, 2014).

John Quackenbush, *Curiosity Guides: The Human Genome* (Imagine: 2014).

Bryan Sykes, *The Seven Daughters of Eve: The Science that Reveals Our Genetic Ancestry* (W. W. Norton & Company, 2002).

John H. Walton, *The Lost World of Adam and Eve: Genesis 2-3 and the Human Origins Debate* (IVP Academic, 2015).

Dan Vogel and Brent Lee Metcalfe, eds., *American Apocrypha: Essays on the Book of Mormon* (Signature Books, 2002).

Chris Stringer, *Lone Survivors: How We Came to Be the Only Humans on Earth* (St. Martin's Griffin, reprint 2013).

Nicolas Wade, *Before the Dawn: Recovering the Lost History of Our Ancestors* (Penguin, 2007).

Bodie Hodge, *Tower of Babel: The Cultural History of Our Ancestors* (Master Books, 2013).

Bill Cooper, *After the Flood: The Early Post-Flood History of Europe Traced Back to Noah* (New Wine Ministries, 1995).

Ken Johnson, *Ancient Post-Flood History: Historical Documents that Point to Biblical Creation* (CreateSpace Independent Publishing Platform, 2010).

Ken Johnson, *Fallen Angels* (CreateSpace Independent Publishing Platform, 2013).

Giovanni Boccaccio, *On the Genealogy of the Gentile Gods* (original publication information unknown, but published somewhere in Italy in 1360).

Francis L. Macrina, *Scientific Integrity: An Introduction Text with Cases*, 2d. ed. (ASM Press, 2000).

Dean Andrew Nicholas, *The Trickster Revisited: Deception as a Motif in the Pentateuch* (Peter Lang, 2009).

Author's Notes: Purposes, Caveats, And Disclaimers

BRIDGING THE GAP FROM ACADEMIA TO MAIN STREET

Before I began this project, I was acutely aware that the world is populated by two basic types of people: regular folks (who compose the great majority of humanity) and academics (who make up a small percentage that has influence vastly out of proportion to the numbers). I have a foot in both worlds; I am surrounded by both kinds of people, and I run in both circles. I knew that bridging the seemingly unbridgeable gap between the two would be one of my biggest challenges in writing this kind of book. The university system in which I teach (Georgia) is fairly typical of other state public systems in America today. The main difference between it and that of, say, California or New York, is that it is located in the Bible Belt. That means it operates within cities and communities where Fundamentalist, Protestant Christianity is the cultural *tour de force*, and where the majority of students come from that background. Even so, a large number of faculty and administrators—yes, even in Georgia—are atheists, agnostics, nominal Christians, spiritualists unaffiliated with any particular religion, or something else. In order to write a book like this, therefore, I knew I had to be impervious to the jeers of my peers; that is, I had to be undaunted by the secular academicians around me who would scoff at any hint that Genesis should be taken as anything other than pure myth. I knew I would be facing the withering criticism of what educated unbelievers and skeptics would likely consider childishly naive Fundamentalism. I had to ask myself, how will I respond to those who laugh at my earnestness and mock my belief? The answer: I will tell them that, "As a scholar, I have questioned my own belief system throughout my adult life. After all, nothing could be more important than knowing the truth about God (if there is a God), which religion is right (if any), and what happens to us when we die (besides a funeral, if anything), right? I don't want to get it wrong! Such a longing to *know*, not just to hope or assume, led me to write a book that tries to answer these hardest of questions." If that answer doesn't satisfy them, I will have done my reasonable duty toward them and can walk away with a clear conscience.

In fairness to the reader, I must confess from the start that, although I am an academician who works in a secular public university system, I am also a believer in God who comes at this topic from a Christian perspective. To be more precise, I grew up in a mainstream Protestant church culture and have been influenced throughout my life mostly by traditional, conservative, orthodox preaching and teaching that has come through a variety of denominations. I am, therefore, what scoffers might derisively call an "Evangelical" and a "Fundamentalist." Let me respond to each of these labels one at a time. I will accept the label of "Evangelical," but I would much prefer to have that term qualified by an adjective such as "Enlightened," "Intellectual," or even "Critical Thinking" Evangelical because that is how I describe myself. True to what Evangelicalism is, I believe spreading the Gospel is the "Great Commission" (meaning top priority) of all believers in Christ. Indeed, it seems nearly impossible to me that one could read the New Testament and not come away with that understanding of the paramount importance of believers being the light of the world, the salt of the earth, a city set on a hill, etc. Whether that means to go door-to-door, accost people on the streets, hold tent revivals, or employ similarly straightforward in-your-face methods is debatable. The old one-liner attributed to St. Francis of Assisi, "Preach the Gospel at all times . . . and use words when necessary," seems about right to me. The only way I can see anyone arriving at any other conclusion about the importance of Evangelicalism within Christianity is to read Revelation 22:11 ("He that is unjust, let him be unjust still . . ." KJV) and decide that it supersedes all that came before it in the New Testament. I believe the application for that verse comes only after an unbeliever has made it clear that he or she rejects the Gospel and refuses to be moved from that position. In such cases, it is neither necessary nor wise to continue trying to persuade him/her. If such folks are to be persuaded, it will be by the Lord's divine intervention, as happened with Saul on the road to Damascus in Acts Chapter 9.

By contrast, I will not accept the label "Fundamentalist." When asked, I always unashamedly answer that I am actually a "*Quasi*-Fundamentalist." After having given a few seconds for the inevitable quizzical look to subside, I go on to explain that Quasi-Fundamentalist means that I believe the Bible is the actual word of God in the sense that it, among all books in the world, uniquely conveys the message of the one true God's existence, his character, and his will and plan for humanity. That does not mean that it is inerrant in the sense that absolute Fundamentalists (Bible Literalists) say it is. The message may indeed be flawed in places, not through the fault of God but rather through the imperfections of the mere mortals involved in writing it down, copying it, and translating it from language-to-language. This includes the ancient Hebrews/Jews who little-by-little over several centuries transcribed the overall message, the church fathers who decided what books to include in the canon, and the translators (which in America means modern English-speaking theologians). When I read the Bible, therefore, I am aware that any given word or sentence may be slightly mistranslated, or that portions of the overall message that any given ancient transcriber wrote might be a misconstruction of what God actually said in revelation to them through the Holy Spirit or through an angelic messenger. "If that is the case," skeptics will surely ask, "then how can anyone know the truth about God? How can any of us know what to believe? And if we can't know for sure what to believe, why should we care? Why not just live and let live, and why not just eat, drink, and be merry until we die?"

My answer: that is where *The Explanation*, comes in. The main goal of this book is to explain to open-minded unbelievers, skeptics, and agnostics (with emphasis on "open-minded") why the overarching message of the Bible is true by showing how it is or at least *can be* believable. To do that, I attempt to make plausible the claims of supernatural, miraculous, and otherwise unbelievable events in the first few chapters of Genesis in light of secular science and other theories to the contrary. My hope is by doing that, some will come to believe, and by believing have their temporal and spiritual lives changed for the better, and, by the butterfly effect, make the world a better place. Making the book of Genesis plausible to committed atheists and those devoted to other religions, however, is admittedly a hard sell, because they have already made up their minds and thus are no longer "open-minded" on the subject. Although I can hope to change the minds of some such readers, I expect most to scoff at my work and to belittle the author. I accept that as par for the course.

Some recent scientific research indicates that each person is born with a genetic predisposition toward either belief in God or atheism. While it may or may not be true that each of us is predisposed one way or another, it is by no means set in stone that any of us must be this way or that. In other words, I believe that any atheist *can* come to believe in God, and any believer *can* lose his or her faith. It seems to me from observation that all committed atheists are what they are for one of two reasons: either 1) as children they were not brought up to believe in God but were taught atheism or some other belief system that doesn't require a God, or 2) they once were believers but eventually stopped believing upon growing older, getting more educated, beginning to ask hard theological questions, becoming increasingly skeptical when they didn't find satisfying answers, developing cynicism about religion in general from having a bad experience with it personally, and often ending up with great contempt for the very idea of there being a God. I doubt that many people in either of these categories will be turned into believers by simply reading *The Explanation*, but everybody is different, and some who are truly seeking answers rather than trying to defend their long-held beliefs, may indeed find their answers, or at least some of them, in this book.

Let me be clear, though: even though I am an Evangelical and would love to win some to Christ through *The Explanation*, I did not write it primarily as a proselytizing tool. I wrote it rather because I am the poster-child for group 2 above, except for the last part about ending up with contempt for God. At some point or other over the five decades of my life, I have passed through every stage in the group 2 series up to that one, but I have never allowed myself to cross over to the dark hopelessness of that last stage. I've been right to the edge where my toes dangled off the side of the precipice, but I have never been able to take the plunge into the abyss. To do so would be to take a leap of faith just as great as that required to believe in God, because all the evidence is not in yet, and therefore the verdict is still out. To make such a decision would therefore be premature and consequently unwise. Scientific and historical knowledge is constantly increasing, and just as much of it seems to point toward the Bible being true as toward a godless universe being true. Each time an astronomer, geologist, archaeologist or such finds something new, it adds to the overall weight of the evidence on one side or the other. Without a doubt, scientists are routinely finding just as much that bolsters the credibility of the Bible as they are finding to tear it down. Likewise, each time an outside-the-box-thinking science fiction writer, cosmologist, or philosopher comes up with a new idea, it must be considered alongside the hard evidence as well. Since this is an ongoing process, I can't close off my mind as if I have already arrived at "the ultimate truth" that waits at the end of the line.

Here scoffers will naturally accuse me of contradicting myself. "If your mind is still open as you claim," they will say, "how can you then throw in with a religious group? How can you call yourself a Christian, much less an Evangelical Christian?" The answer is that I apply a version of Pascal's Wager that goes like this: *If I'm wrong, and there is no God, then when I die, what I believed and how I lived never mattered anyway. If I'm right, though, and there is a God, then when I die, I will like my chances a lot better if I've lived a moral life and tried to promote morality, tolerance, love, and faith.* The scoffer's reply to that philosophy is likely to be, "What if there is a God, after all, but it's not the Christian God?" Answer: I'm convinced it is the Christian God. Why? Because of all I have written in this book concerning what really happened in the beginning. It goes back to the fact that, while standing there on the edge again and again in my late 30's and early 40's peering off into the blackness of the abyss, I eventually came to decide that just because I had not yet found satisfying answers to the hard questions didn't mean they weren't out there. If they were out there, I said, I must find them.

I realized that some greater thinker than I may have already found and published them, and I just hadn't discovered that person or his/her work yet. Or, perhaps no one has yet found and published them, and if that is the case, would it not be my duty to find as many of them as I could and publish them? Would I not be abdicating my responsibility as an open-minded critical thinker if I did anything less? So I challenged myself thusly, and I accepted the challenge. What I have found after many years of searching, therefore, is that

Christian theology seasoned with the right amount of secular science and out-of-the-box-thinking allows for the best, if not the only reasonable, explanation for the mysteries of life and the universe. It provides satisfying answers, in other words, to more of the hard questions than anything else does, at least in my opinion. I have attempted to show that in *The Explanation*.

BEING WISE AS A SERPENT BUT HARMLESS AS A DOVE

One main reason for writing *The Explanation* comes from the biblical principle stated in I Corinthians 14:20, where the apostle Paul says, "Brethren, be not children in understanding: howbeit in malice be ye children, but in understanding be men." Similarly, in Matthew 10:16, Jesus told his disciples, "Behold, I send you forth as sheep in the midst of wolves: be ye therefore wise as serpents, and harmless as doves." It seems to me that true Christians have always been like sheep in the midst of wolves, as we still are today. The Bible contains no scripture in which either Christ or any of the apostles ever instructed us to turn our minds off or to check our brains at the door before we enter the church. Instead, the Bible is full of examples where God used educated men and/or wise men to accomplish his purposes. The fact that Christ chose some uneducated fishermen as disciples just means He *can* and *will* use the lowly, and that God is no respecter of persons [Acts 10:34, James 2:9]. However, in most cases God's chosen instruments were at least literate if not well-educated in the scriptures, Jewish traditions, and/or Roman law. Paul is a perfect example of one who was all of those things. Is it any wonder, then, that he penned more of the New Testament than anyone else?

Christ himself was clearly literate, as He read aloud from the scroll of Isaiah in the Temple [Luke 4:16-20]. He was likewise full of understanding and wisdom, whether from formal education, supernatural revelation, or divine inheritance, as He first displayed at the age of 12 in the temple [Luke 2:42-47]. Either way, He displayed a complexity of thought about Jewish law and customs that no one alive at the time had ever seen. If nothing else, his parables alone prove him to be among the wisest people to have ever walked the earth. We must also recognize that He had the ability to read the scriptures and see himself as the Messiah in them, to see the nature of his earthly ministry in them, and foresee his own death on the cross in them. Again, He had the benefit of being the son of God—God incarnate—so we cannot disregard the possibility of divine inheritance in all of this, but regardless of where He got his knowledge, there is no question that He was full of it, and that He used it wisely.

The complication that we Christians often run into on this subject is that the most well-educated among us in theology tend to have one of two problems, if not both: either our book-learning inflates our ego or else it causes us to lose our childlike faith. Likewise, the less-educated among us tend to have one of two problems, if not both: either we have an inferiority complex about our lack of education or else we have a foolish arrogance that, despite our lack of "formal" education, we know the Bible just as well as any ivory tower theologian or big name preacher. The thinking behind the first three of these complications is self-explanatory, but the last requires elaboration. As one who has lived a great deal of my life in this mode of thought, I can speak with some authority on it. Although it is mostly subconscious, it goes something like this:

God loves poor little ol' me just as much as any uppity seminary-trained preacher, because Christ died for all of us. What I lack in the way of book-learning, I know from my personal relationship with the Lord. I have childlike faith, just like Jesus told us all to have. That's better than book-learning, anyway. Besides, I've heard that a lot of those Bible colleges explain away the scriptures. They corrupt the students and make them question the faith. They've traded in the true Gospel for a bunch of head knowledge. If I were to go and get myself educated, I would be doing the same thing—trading in my heartfelt faith for head knowledge about the Bible.

For many Christians, this actually may turn out to be true, like a self-fulfilling prophecy. Education in theology can, and indeed does, cause many to fall away from the faith. It destroys the simple-minded faith of some by shattering the Sunday School version of Bible stories that we all learned as children. (Of course, secular education in the sciences and social sciences has the same effect on many Christians, but that's a different issue.) It replaces simplicity with complexity, in other words, and many of us don't know how to handle that. We wrestle with the scriptures and struggle to understand until we grow weary of the battle raging in our minds. The result is often not that we backslide into sin but that we take a deliberate swan dive into unbelief, and from there we fall into a deep pit of debauchery and depravity.

There is hope, however. Like so many prodigal sons [Luke Chapter 15] and daughters, once we hit bottom and are engulfed in utter despair, then and perhaps only then can we see the folly of what we have done in leaving our loving Father's house to try to find happiness and fulfillment on our own. Indeed, standing in the pig pen up to our knees in the putrid slop of a life spent floundering in apathy, confusion, or unbelief, we look for some way to recapture our personal dignity only to find none. We then long for the good ol' days when, despite whatever rules or expectations we were bound by at home, at least somebody loved us there and provided for our most basic material needs. And as we weigh our options, we can see clearly that returning to the Father and returning to faith not only offers the hope of eternal life, but it actually guarantees a better quality of life in the here-and-now.

This scenario is of course not the only one that may play out in any given individual's life, but it is not an uncommon one. Not all people who have suffered their faith to be shipwrecked through book-learning have the happy ending of the prodigal son. For them, it would have been better to remain uneducated and preserve the simplicity of thought that went hand-in-hand with their childlike faith. For that reason, I caution readers who have childlike faith and prefer simplicity to complexity to count the cost [Luke 14: 28-32] before reading books like this one.

Acknowledgements

I am grateful to Dr. H. Lee Cheek, Dean of the School of Social Sciences at East Georgia State College, for supporting my work on this project. His moral support, along with the financial support he helped arrange through a departmental summer research grant, made completion of this project much easier than it otherwise would have been. Thanks likewise go to Elizabeth Gilmer and the East Georgia State College Foundation for providing funds for that grant.

I also appreciate the support of other senior administrators of East Georgia State College, including President Bob Boehmer and Vice-President of Academic Affairs Tim Goodman. Working with such good people makes my job much more than just a means to a paycheck; it makes my job a pleasant and dynamic experience.

About The Author

T. Adams ("Tommy") Upchurch was born in Lexington, Mississippi, on the night of Lyndon Johnson's presidential election in 1964. Moving to the neighboring town of Durant as a child, he grew up a Southern Baptist and gave his life to Christ at age 10. Tommy attended a black majority public school in a time and place filled with racial tension. After working in various occupations and living in sundry locations, he matriculated at Holmes Community College in Goodman, Mississippi, then earned undergraduate and graduate degrees in the field of Education at Delta State University in Cleveland, Mississippi. In 2001, he earned his Ph.D. in History at Mississippi State University.

Dr. Upchurch teaches History at East Georgia State College in Statesboro, Georgia, where he is now a tenured full-professor. Over the years, he has authored many scholarly books, journal articles, and encyclopedia entries. Among his most notable publications are *Christian Nation? The United States in Popular Perception and Historical Reality* (Praeger, 2010) and *Legislating Racism: The Billion Dollar Congress and the Birth of Jim Crow* (University Press of Kentucky, 2004).

Married to Dr. Linda Liles Upchurch, he has three children and, so far, two grandchildren. He and his wife are members of Compassion Christian Church in Statesboro, Georgia.

www.ingramcontent.com/pod-product-compliance
Lightning Source LLC
Chambersburg PA
CBHW032015230426
43671CB00005B/94